Chemical Calculations at a Glance

Chemical Calculations at a Glance

Paul Yates

Blackwell
Publishing

Editorial offices:
Blackwell Publishing Ltd, 9600 Garsington Road, Oxford OX4 2DQ, UK
Tel: +44 (0)1865 776868
Blackwell Publishing Inc., 350 Main Street, Malden, MA 02148-5020, USA
Tel: +1 781 388 8250
Blackwell Publishing Asia Pty Ltd, 550 Swanston Street, Carlton, Victoria 3053, Australia
Tel: +61 (0)3 8359 1011

First published 2005 by Blackwell Publishing Ltd

Library of Congress Cataloging-in-Publication Data
Yates, Paul.
Chemical calculations at a glance / Paul Yates. – 1st ed.
p. cm.
Includes index.
ISBN 1-4051-1871-7 (pbk. : alk. paper)
1. Chemistry – Mathematics. I. Title.

QD39.3.M3Y382 2005
540′.1′51 – dc22 2004019623

ISBN 1-4051-1871-7

A catalogue record for this title is available from the British Library

Set in 10/11 pt Times
by TechBooks, New Delhi, India
Printed and bound in India
by Replika Press Pvt. Ltd, Kundli

The publisher's policy is to use permanent paper from mills that operate a sustainable forestry policy, and which has been manufactured from pulp processed using acid-free and elementary chlorine-free practices. Furthermore, the publisher ensures that the text paper and cover board used have met acceptable environmental accreditation standards.

For further information on Blackwell Publishing, visit our website:
www.blackwellpublishing.com

Contents

Preface

Why this book? I realise that I probably only have a few seconds to answer this question before you decide whether to read any further, so here are two points which summarise the philosophy behind it.

The first is that wherever possible the mathematics is presented through the use of actual chemical examples. I firmly believe that this is much more valuable than using abstract problems. The end of chapter questions all contain traditional problems using x and y, but then go on to explore chemistry using the wealth of symbols to be found in physical chemistry textbooks.

Second, at the end of the book you will find outline solutions to all the problems, rather than just the final answers. Again, I believe this is vital for making this a book that can be used by students, whether or not an instructor is present.

I have aimed this book at the student with GCSE mathematics who has embarked upon a degree in chemistry. The mathematics required, particularly in physical chemistry, may seem daunting. My intention is to provide a reference that can be used for assistance at any point during the chemistry degree, but particularly in the first year.

In places I have gone for a clear workable approach, rather than being mathematically rigorous. I hope that any mathematicians who happen to read this book will forgive me for this.

Acknowledgements

I would like to thank Paul Sayer at Blackwell Publishers for suggesting that I start this project, for his encouragement and his faith in my ability to complete the work. Former students in the School of Chemistry and Physics at Keele University gave me much inspiration for the material contained in this book and convinced me of the value of this approach. Finally thanks go to Julie, Catherine and Christopher for their support and encouragement and constant belief in me.

Paul Yates
Keele
July 2004

Introductory Skills

1. Rules of Indices

An index (plural indices) is simply the power to which a number is raised. For example, x^2 is said to have an index of 2; the index of x^6 is 6. When numbers involving indices to the same base are combined, a set of simple rules is obeyed as follows:

$$x^m \times x^n = x^{m+n} \quad \text{for example, } x^3 \times x^2 = x^{3+2} = x^5 \quad \text{or} \quad 3^4 \times 3^5 = 3^{4+5} = 3^9 = 19\,683$$

$$\frac{x^m}{x^n} = x^{m-n} \quad \text{for example, } \frac{x^6}{x^4} = x^{6-4} = x^2 \quad \text{or} \quad \frac{8^4}{8^3} = 8^{4-3} = 8^1 = 8$$

$$\left(x^m\right)^n = x^{mn} \quad \text{for example, } \left(x^2\right)^3 = x^{2\times3} = x^6 \quad \text{or} \quad \left(2^3\right)^2 = 2^{3\times2} = 2^6 = 64$$

$$x^0 = 1 \quad \text{for example, } a^0 = 1 \quad \text{or} \quad 5^0 = 1$$

$$\frac{1}{x^n} = x^{-n} \quad \text{for example, } \frac{1}{x^3} = x^{-3} \quad \text{or} \quad \frac{1}{3^2} = 3^{-2}$$

$$x^{1/n} = \sqrt[n]{x} \quad \text{for example, } x^{1/2} = \sqrt{x} \quad \text{or} \quad x^{1/3} = \sqrt[3]{x}$$

As well as the negative index shown above, it is also possible to have fractional indices. Terms such as $x^{5/2}$, $x^{-3.6}$ and $3^{4.2}$ are valid.

Expressions involving indices can be evaluated on a calculator using the x^y key. For example, to calculate 3.4^3 you would input 3.4, then press x^y, input 3, and then press the = key to give 39.304.

Reaction between Hydrogen and Iodine

Indices frequently occur in chemistry in expressions for equilibrium and rate constants that involve concentrations. The concentration of a species is normally shown in square brackets, such as [HCl], and is expressed in units of mol dm^{-3}.

For the chemical reaction between hydrogen gas and solid iodine to give hydrogen iodide, represented by the reaction

$$H_2 + I_2 \rightleftharpoons 2\,HI$$

the equilibrium constant K is given by the expression

$$K = \frac{[HI]^2}{[H_2][I_2]}$$

In the special case where $[H_2] = [I_2]$, the expression becomes

$$K = \frac{[HI]^2}{[H_2][H_2]} = \frac{[HI]^2}{[H_2]^2} = [HI]^2[H_2]^{-2}$$

The rate of the forward reaction in the above equilibrium is given by

$$Rate = k[H_2][I_2]$$

This is said to be first order in H_2 and first order in I_2. The reverse rate is given by

$$Rate = k[HI]^2$$

and is said to be second order in HI. Thus in a rate equation, the order of each reactant is simply the index of its concentration in the expression. The overall order is simply the sum of all the indices in the expression, in this case 2 for both forward and reverse reactions.

Questions

1. Simplify the following expressions involving indices as far as possible:
   (a) $a^3 \times a^5$
   (b) $x^2 \times x^6$
   (c) $y^4 \times y^3$
   (d) b^6/b^3
   (e) x^{10}/x^2

2. Simplify the following expressions involving indices as far as possible:
   (a) $(c^4)^3$
   (b) z^0
   (c) $1/y^4$
   (d) $x^5 \times x^{-5}$
   (e) x^{-2}/x^{-3}

3. Give the order with respect to each reactant, and the overall order, in each of the following rate equations:
   (a) $Rate = k[CH_3CHO]^{3/2}$
   (b) $Rate = k[BrO_3^-][Br^-][H^+]^2$
   (c) $Rate = k[NO]^2[Cl_2]$

4. Simplify the rate equation

$$Rate = k[H_2O_2][H^+][Br^-]$$

   in the case when $[H^+] = [Br^-]$.

5. The acid dissociation constant for the equilibrium

$$H_2CO_{3(aq)} \rightleftharpoons H^+_{(aq)} + HCO_3^-{}_{(aq)}$$

   is given by

$$K_a = \frac{[H^+][HCO_3^-]}{[H_2CO_3]}$$

   Simplify this expression by considering the relationship between $[H^+]$ and $[HCO_3^-]$.

2. Scientific Notation

Scientific notation is frequently used in chemistry where we encounter both very small and very large quantities. Examples include a wavelength of 1.54×10^{-10} m and the value of Planck's constant 6.63×10^{-34} J s. The common feature of these values is that they consist of a number between 1 and 10 multiplied by a power of 10. This is true of any number expressed in scientific notation.

Powers of 10

The usual notation for powers of 10 is:

$$\begin{aligned} 10^1 &= 10 \\ 10^2 &= 10 \times 10 = 100 \\ 10^3 &= 10 \times 10 \times 10 = 1000 \\ 10^4 &= 10 \times 10 \times 10 \times 10 = 10\,000 \end{aligned}$$

and so on. In general, 10^n is equal to the number written as 1 followed by n zeros.

The value of n may also be negative. For example, $10^{-n} = 1/10^n$, as shown in Chapter 1. In this case we have

$$\begin{aligned} 10^{-1} &= 0.1 \\ 10^{-2} &= 0.01 \\ 10^{-3} &= 0.001 \\ 10^{-4} &= 0.0001 \end{aligned}$$

and so on. In general, 10^{-n} is equal to the number written as 1 in the nth decimal place.

Interpreting scientific notation

It follows from the above that the number written as

$$3.67 \times 10^3 = 3.67 \times 1000 = 3670$$

and that

$$8.382 \times 10^6 = 8.382 \times 1\,000\,000 = 8\,382\,000$$

Also

$$2.9 \times 10^{-2} = 2.9 \times 0.01 = 0.029$$

and

$$6.397 \times 10^{-7} = 6.397 \times 0.000\,000\,1 = 0.000\,000\,639\,7$$

Converting to scientific notation

Suppose we have the number 8352. First we need to write the appropriate number between 1 and 10, which is 8.352. This number needs to be multiplied by 1000 or 10^3 to give the original value. So in scientific notation we write this as 8.352×10^3.

Now consider 0.000 004 39. The appropriate number between 1 and 10 is 4.39. To give the original answer we need to multiply by 0.000 001 or 10^{-6}. So in scientific notation this is given as 4.39×10^{-6}.

A quick way of making such transitions is to note that in the first case the decimal point has moved 3 places to the left, so the appropriate power of 10 is 3. In the second, the decimal point has moved 6 places to the right, so the appropriate power of 10 is -6.

Chemical examples

The speed of light in a vacuum can be quoted as $2.997\,924\,58 \times 10^8$ m s^{-1}. This can be evaluated as $2.997\,924\,58 \times 100\,000\,000$ m s^{-1}, which gives 299 792 458 m s^{-1}.

The helium spectrum contains a line with a wavelength of 2058.2 nm. This can be written as 2058.2×10^{-9} m (see Appendix 1), or $2058.2 \times 0.000\,000\,001$ m, which equates to 0.000 002 058 2 m.

The viscosity of carbon dioxide at 575.15 K has been measured as 0.000 268 2 p. In scientific notation we would write this as 2.682 p, realising that we need to include a factor of 0.0001 or 10^{-4} to correctly scale this. The value could thus be quoted as 2.682×10^{-4} p.

A wavenumber spacing of 1 cm^{-1} is equivalent to a difference in energy of 0.000 123 984 eV. This can be written as $1.239\,84 \times 0.0001$ eV or $1.239\,84 \times 10^{-4}$ eV.

Questions

1. Write the following physical constants without using powers of 10:
   (a) $F = 9.649 \times 10^4$ C mol^{-1}
   (b) $R = 1.097 \times 10^7$ m^{-1}
   (c) $\mu_o = 12.57 \times 10^{-7}$ N A^{-2}
   (d) $a_o = 5.292 \times 10^{-11}$ m
   (e) $h = 6.626 \times 10^{-34}$ J s

2. Rewrite the following physical constants in scientific notation:
   (a) $e = 0.1602 \times 10^{-18}$ C
   (b) $E_h = 4360 \times 10^{-22}$ J
   (c) $m_e = 0.009\,109 \times 10^{-28}$ kg
   (d) $N_A = 602.2 \times 10^{21}$ mol^{-1}
   (e) $R = 0.000\,831\,4 \times 10^4$ J K^{-1} mol^{-1}

3. Write the following quantities without using powers of 10:
   (a) 9.4×10^{-5} bar
   (b) 3.72×10^{-2} cm
   (c) 1.8×10^{-4} MHz
   (d) 1.95×10^{-3} kJ mol^{-1}
   (e) 7.19×10^4 s^{-1}

4. Write the following quantities using scientific notation:
   (a) 0.0417 nm
   (b) 352 s
   (c) 2519 m s^{-1}
   (d) 0.076 kJ mol^{-1}
   (e) 579 eV

3. Units

Units are one of the most important, yet frequently forgotten, elements of physical quantities. Typically in physical chemistry we are concerned with measurement; we might talk about a mass of 5.72 g, a concentration of 0.15 mol dm^{-3}, or a wavelength of 560 nm. These quantities are very different from a mass of 5.72 kg, a concentration of 0.15 mmol dm^{-3}, or a wavelength of 560 pm, even though the same number is involved.

Physical quantities

A physical quantity, in other words something that we can measure, is made up of a number combined with its unit. Normally in chemistry these units are taken from the Système International, know as SI for short. Appendix 2 gives the commonly used SI units in chemistry in terms of these base units.

Prefixes

A unit may be prefixed with one of the symbols shown in Appendix 1, each of which represents a different power of 10. This allows us to concisely represent quantities of widely different magnitudes. For example, the length 0.154 nm is the same as 0.154×10^{-9} m; this is the same as 1.54×10^{-10} m.

Converting units

Occasionally SI units are not convenient, or we are working in an area which has traditionally used a different set of units. Then we need to be able to convert between different sets of units. To do this we simply use the appropriate conversion factors and apply the standard rules of mathematics. Some of these units are given in Appendix 3.

Combining units

If we need to calculate a quantity by combining two other quantities, we simply treat the units in the same way as their associated numbers.

Tables and graphs

Quantities which appear as table headings or axis labels on graphs are represented with the symbol for that quantity divided by its unit, for example, V/cm^3. This then allows the entry in the table or the point on the graph to be read as a pure number.

The gas constant

This constant, symbol R, is normally expressed as 8.314 J K^{-1} mol^{-1}. However, there are occasions when it is easier to use when expressed in units of dm^3 atm K^{-1} mol^{-1}. Comparison of the units shows that we need to find the conversion factor between J and dm^3 atm, as the units K^{-1} mol^{-1} are the same in both cases. Start by establishing that:

$$1\ \text{atm} = 101\,325\ \text{Pa} \quad \text{(Appendix 3)}$$
$$1\ \text{Pa} = 1\ \text{N m}^{-2} \quad \text{(Appendix 2)}$$
$$1\ \text{J} = 1\ \text{N m} \quad \text{(Appendix 2)}$$

and also that

$$1\ \text{dm} = 10^{-1}\ \text{m} \quad \text{(Appendix 1)}$$
$$1\ \text{dm}^3 = (10^{-1}\ \text{m})^3$$
$$= 10^{-3}\ \text{m}^3$$

Successive substitutions give:

$$\begin{aligned} R &= 8.314\,\mathrm{J\,K^{-1}\,mol^{-1}} \\ &= 8.314\,\mathrm{N\,m\,K^{-1}\,mol^{-1}} \\ &= 8.314\,\mathrm{(Pa\,m^2)\,m\,K^{-1}\,mol^{-1}} \\ &= 8.314\,\mathrm{(atm/101\,325)\,(m^2)\,m\,K^{-1}\,mol^{-1}} \\ &= 8.314\,\mathrm{(atm/101\,325)\,m^3\,K^{-1}\,mol^{-1}} \\ &= 8.314\,\mathrm{(atm/101\,325)\,10^3\,dm^3\,K^{-1}\,mol^{-1}} \\ &= 0.082\,06\,\mathrm{dm^3\,atm\,K^{-1}\,mol^{-1}} \end{aligned}$$

Calculating a wavelength

The equation to calculate the wavelength λ associated with a particle of mass m moving with a velocity v is given by

$$\lambda = h/mv$$

The quantity h is Planck's constant, 6.63×10^{-34} J s. This and other physical constants are listed in Appendix 4.

Suppose that an electron of mass $m = 9.11 \times 10^{-31}$ kg has velocity $v = 2.42 \times 10^6$ m s^{-1}, then

$$\lambda = \frac{6.63 \times 10^{-34}\,\mathrm{J\,s}}{9.11 \times 10^{-31}\,\mathrm{kg} \times 2.42 \times 10^6\,\mathrm{m\,s^{-1}}}$$

From Appendix 2, 1 J = 1 N m = 1 kg m^2 s^{-2}, so that

$$\lambda = \frac{6.63 \times 10^{-34}\,\mathrm{kg\,m^2\,s^{-2}\,s}}{9.11 \times 10^{-31}\,\mathrm{kg} \times 2.42 \times 10^6\,\mathrm{m\,s^{-1}}}$$

The numerical part is easily evaluated using a calculator, while most of the units cancel to give $\lambda = 3.01 \times 10^{-10}$ m, which has the units we would expect for wavelength.

Boiling points and enthalpies of vaporisation

The following table shows the boiling point, T_b, of benzene and trichloromethane, along with their enthalpies of vaporisation:

	T_b/K	$\Delta_{vap}H$/kJ mol^{-1}
C_6H_6	353	30.8
$CHCl_3$	334	29.4

To interpret such a table, simply equate a table entry with its column heading. For example, for $CHCl_3$ the table gives

$$\Delta_{vap}H/\mathrm{kJ\,mol^{-1}} = 29.4$$

Multiplying both sides of this equation by the units kJ mol^{-1} gives $\Delta_{vap}H = 29.4$ kJ mol^{-1}.

Kinetics of N_2O_5 decomposition

The graph (Fig. 3.1) has the slight complication that the $[N_2O_5]$ values are all multiplied by a constant. However, the procedure is just the same as in the table above, so for the fourth point from the left

$$[\mathrm{N_2O_5}]/10^{-3}\ \mathrm{mol\,dm^{-3}} = 360$$

i.e. the axis label equal to the pure number represented by the point. Then by multiplying both sides by 10^{-3} mol dm^{-3}

$$\begin{aligned} [\mathrm{N_2O_5}] &= 360 \times 10^{-3}\,\mathrm{mol\,dm^{-3}} \\ &= 0.360\,\mathrm{mol\,dm^{-3}} \end{aligned}$$

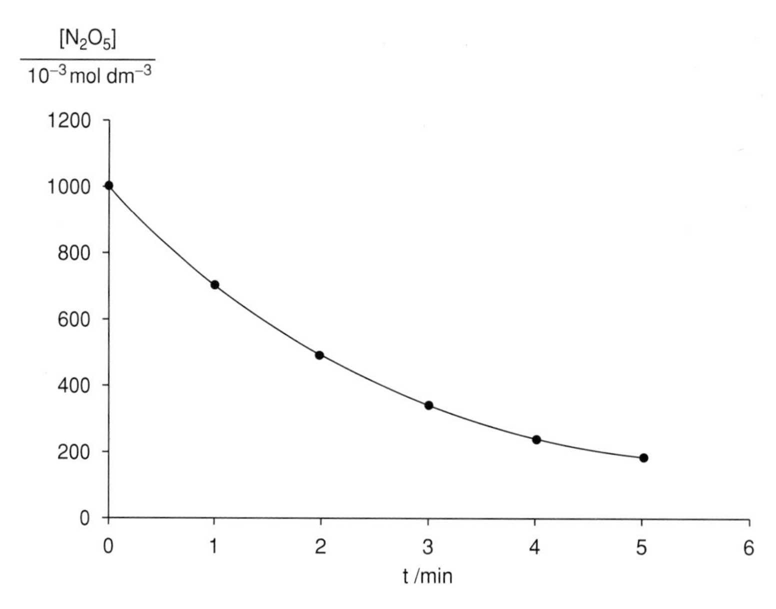

Figure 3.1 Graph of N_2O_5 concentration $[N_2O_5]$ against time t during decomposition

Questions

To successfully answer the questions in this chapter you will also have to make use of the information contained in Appendices 1, 2 and 3.

1. Write the following quantities using powers of 10:
   (a) 54.2 mg
   (b) 1.47 aJ
   (c) 3.62 MW
   (d) 4.18 kJ mol^{-1}
   (e) 589 nm

2. Write the following quantities using appropriate prefixes:
   (a) 3.0×10^8 m s^{-1}
   (b) 101 325 Pa
   (c) 1.543×10^{-10} m
   (d) 1.68×10^2 kg m^{-3}
   (e) 7.216×10^{-4} mol dm^{-3}

3. Convert the following quantities to the units shown:
   (a) 1.082 g cm^{-3} to kg m^{-3}
   (b) 135 kPa to N cm^{-2} (1 Pa = 1 N m^{-2})
   (c) 5.03 mmol dm^{-3} to mol m^{-3}
   (d) 9.81 m s^{-2} to cm ms^{-2}
   (e) 1.47 kJ mol^{-1} to J $mmol^{-1}$

4. Convert the following quantities to the SI units shown:
   (a) 4.28 Å to pm
   (b) 54.71 kcal to kJ
   (c) 3.6 atm to kPa
   (d) $2.91E_h$ to J
   (e) $3.21a_0$ to nm

5. Express the results of the following calculations in appropriate SI units:
   (a) 5.62 g × 4.19 m s^{-2}
   (b) 4.31 kN/10.46 m^2
   (c) 2.118×10^{-3}J/3.119×10^{-8} C
   (d) 6.63×10^{-34} J s × 3×10^8 m s^{-1}/909 nm
   (e) 4.16×10^3 Pa × 2.14×10^{-2} m^3

6. The following pairs of labels could be used in either a table or on a graph. In each case state how the label should be changed (if at all) in order to represent the data correctly.
   (a) $p(\text{Pa})$; $T(\text{K})$
   (b) t/sec; $c/\text{mol dm}^{-3}$
   (c) $c/\text{mol per dm}^3$; $\rho/\text{g cm}^{-3}$
   (d) $\frac{1}{c/\text{mol dm}^{-3}}$; t/s
   (e) $\frac{n_i}{n_j}$; $T/\text{K}(\times 10^3)$

4. Significant Figures and Decimal Places

In reality, instead of significant figures and decimal places we should talk about significant digits and the number of digits after the decimal point. It is not surprising that these two terms are often confused, as it is possible for the number of significant digits to be the same as the number of decimal places. Both are also related to the precision with which a number is quoted.

Decimal places

The number of digits after the decimal point is determined simply by counting them. For example, the number 478.32 is given to two decimal places.

Significant figures

The number of significant figures is a harder quantity to determine, but we can start by applying some simple rules.

1. Non-zero digits are always significant.
2. Leading zeros are never significant. For example, the number 0.005 32 contains three significant digits.
3. 'Enclosed' zeros are always significant. For example, the number 507.03 is quoted to five significant figures.

More problematic is the case of trailing zeros. For example, we may not know whether the number 500 represents 500 exactly, or is 502 rounded to the nearest 10, or 521 rounded to the nearest hundred. We have to consider each case individually, depending on what other information is available and using our discretion. If, however, the trailing zeros appear after a decimal point they are significant – because omitting them only changes the precision with which the number is quoted.

Rounding

Sometimes a number has to be quoted to a specified number of decimal places or significant figures. This often involves truncating or rounding that number. The basic rule here is to take the number of digits required, filling out with zeros as required, then to look at the next digit. If it is 5 or greater the digit before it increases by 1, otherwise it remains as it is. If the digit to be increased is 9, it becomes 0 and the digit immediately to the left increases by 1.

This will be clearer once a few examples are given:

(a) 522.842 to 1 decimal place (or 4 significant figures) is 522.8. The next digit is 4, so no rounding up is required.
(b) 7.289 to 2 decimal places (or 3 significant figures) is 7.29. We truncate to 7.28, but the next digit is 9 so rounding up is required.
(c) 532 to 2 significant figures is 530. We give the two figures, fill out with zeros to 530, then look at the next digit, which is 2, so no rounding up is required.
(d) 6187 to 3 significant figures is 6190. We give the three figures, fill out with zeros to 6180, then look at the next digit which is 7 so rounding up is required.
(e) 69.8 to 2 significant figures is 70. We give the two figures, then look at the next digit which is 8. We then round the 9 to 0 and increase the digit immediately to the left, so 6 becomes 7.

Typical chemical quantities

The following are some examples of typical chemical quantities that have been rounded:

(a) A density of 0.7248 g cm^{-3} is given to 4 decimal places and 4 significant figures.
(b) An enthalpy change of 6024 J mol^{-1} is given to 0 decimal places and 4 significant figures.
(c) A standard electrode potential of 0.042 V is given to 3 decimal places and 2 significant figures.
(d) A concentration of 0.250 mol dm^{-3} is given to 3 decimal places and 3 significant figures. Here the trailing zero is significant as it would have been omitted if the concentration had been determined less precisely as 0.25 mol dm^{-3}.
(e) In the statement 'most chemical reactions have activation enthalpies between 40 and 400 kJ mol^{-1}' both quantities are given to 1 significant figure. The context implies that these are rounded values, so the trailing zeros are not significant.
(f) One line in the sodium doublet occurs at a wavelength of 589.76 nm, given to 5 significant figures. Using the rules, to 2 significant figures this would be 590 nm; to 3 significant figures it would also be 590 nm.
(g) We are told that a spectrum has been taken at 300 K. From this we cannot deduce the number of significant figures in the temperature as we do not know how precisely this has been determined. The number of significant figures could be 1, 2 or 3.

Questions

1. Give each of the following numbers to (i) 3 significant figures (ii) 2 decimal places:
   (a) 41.62
   (b) 3.959 29
   (c) 10 004.91
   (d) 0.007 16
   (e) 0.9997

2. Give each of the following quantities to (i) 4 significant figures (ii) 1 decimal place:
   (a) 589.929 nm
   (b) 103.14 kJ
   (c) 0.100 46 mol dm^{-3}
   (d) 32.8479 ms
   (e) 101 325 Pa

3. Give the number of significant figures in each of the following numbers:
   (a) 1.497
   (b) 1.0062
   (c) 0.000 791 24
   (d) 1.500
   (e) 64.0020

4. Give the number of significant figures in each of the following quantities:
   (a) 432.1 kJ mol^{-1}
   (b) 909.1 nm
   (c) 0.003 52 m s^{-1}
   (d) 4.730 g
   (e) 0.082 06 dm^3 atm K^{-1} mol^{-1}

SECTION

B

Handling Data

5. Calculated Quantities

If we perform a calculation such as 19.24×56.32, our calculator displays the result as 1083.5968. As we saw earlier, the number of digits displayed is an indication of the precision of a number. We may suspect that the calculator is giving us more digits than are justified. How then, do we decide how many digits to quote?

Sums and differences

Such calculations are best considered by writing the example by hand in the form we may have first used to do such operations. For example,

$$\begin{array}{r} 3.43 \\ \underline{+16.2} \\ \underline{} \end{array}$$

You may remember that such summations are worked out from right to left, by adding the digits in each column in turn. If we look at the far right column, we see that in the top number we have a 3, while below it is a blank. It might be tempting to think of this blank as a zero, but in fact the blank is telling us that the digit in that position is undefined.

Consequently, the summation in that column is really 3 plus an undefined number, which itself is undefined. The result of this should thus be written as a blank. If we do this, the problem doesn't arise elsewhere and so the result of the summation should be written as 19.6.

In practice, we apply this rule by remembering that for sums and differences the answer should be quoted to the same number of decimal places as that in the least well defined piece of data. For example,

the result of $(3.95732 - 2.36)$ should be given to 2 decimal places

the result of $(8.76 + 7)$ should be given to 0 decimal places

Also we need to round as appropriate, by considering one digit further than given in the answer. So, for example, $(3.69 - 2.1)$ gives 1.59, which we then round to 1.6 to give the appropriate level of precision.

Products and quotients

The rule here is to quote the answer to the same number of significant figures as that in the least precise piece of data. For example,

the result of (3.14×2.9) should be given to 2 significant figures

the result of $(0.001\,643/1.44)$ should be given to 3 significant figures

As before, each answer should be rounded if appropriate, i.e. in 1.92×3.087, which gives 5.927 04, is rounded to 5.93 when given to 3 significant figures.

Examples in chemistry

(a) The molecular mass of hydrogen chloride is found by adding the atomic masses of hydrogen and chlorine. This gives:

$$M_r(\mathrm{HCl}) = (1.007\,94 + 34.453)\,\mathrm{g\,mol^{-1}}$$

using published data. The answer needs to be given to 3 decimal places, and so is 35.461 g mol^{-1} once the appropriate rounding takes place.

(b) The electromotive force $E^{\ominus}$ produced by the silver/silver chloride cell is found by taking the difference in potential of the constituent cells. We thus have:

$$\begin{aligned} E^{\ominus} &= E^{\ominus}(Ag^{+}|Ag) - E^{\ominus}(AgCl|Ag) \\ &= 0.7996\ \text{V} - 0.222\,33\ \text{V} \\ &= 0.5773\ \text{V} \end{aligned}$$

which is given to 4 decimal places since this is the precision with which $E^{\ominus}(Ag^{+}|Ag)$ is given. Note that the answer is also rounded up.

(c) Trouton's rule states that, for an undissociated liquid, the value of the entropy of vaporisation $\Delta_{vap}S$ is approximately 88 J K^{-1} mol^{-1}. Since $\Delta_{vap}S$ is equal to $\Delta_{vap}H/T_b$, the enthalpy of vaporisation can be found from the formula

$$\Delta_{vap}H = \Delta_{vap}S \times T_b$$

where T_b is the absolute temperature of boiling

For toluene this calculation gives

$$\begin{aligned} \Delta_{vap}H &= 88\ \text{J K}^{-1}\text{mol}^{-1} \times 383.77\ \text{K} \\ &= 34\,000\ \text{J mol}^{-1} \qquad \text{or} \qquad 34\ \text{kJmol}^{-1} \end{aligned}$$

Note that the answer can only be given to 2 significant figures, and in this case, if the answer is expressed in J mol^{-1}, the number has to be padded out with zeros.

(d) Glucose-6-phosphate isomerises to fructose-6-phosphate. The equilibrium constant K for this reaction can be calculated as the ratio of the two concentrations, as in the expression:

$$K = \frac{0.013\ \text{mol dm}^{-3}}{0.0249\ \text{mol dm}^{-3}}$$

This gives a value for K of 0.52, as only 2 significant figures are justified from the first of these quantities. No rounding is required in this case.

Questions

1. Give the results of each of the following calculations to the appropriate level of precision:
   (a) $1.092 + 2.43$
   (b) $6.2468 - 1.3$
   (c) $100 + 9.1$
   (d) 42.8×36.194
   (e) $2.107/32$

2. Give the results of each of the following calculations involving physical quantities to the appropriate level of precision:
   (a) 9.021 g/10.7 cm^3
   (b) 104.6 kJ mol^{-1} + 98.14 kJ mol^{-1}
   (c) 1.46 mol/12.2994 dm^3
   (d) 3.61 kg × 2.1472 m s^{-1}
   (e) 3.2976 g − 0.004 g

3. The volume V of an ideal gas is given by the equation

$$V = \frac{nRT}{p}$$

where n is the amount of gas, T its absolute temperature and p the pressure. R is the gas constant.

Calculate the volume, quoted to an appropriate degree of precision, of 2.42 moles of an ideal gas at 295 K and 52.47 kPa pressure, taking R as
   (a) 8.3 J K^{-1} mol^{-1}
   (b) 8.31 J K^{-1} mol^{-1}
   (c) 8.314 J K^{-1} mol^{-1}

4. Aluminium chloride, which is used as an industrial catalyst, can be prepared by the reaction

$$2\,Al_{(s)} + 6\,HCl_{(g)} \rightarrow 2AlCl_{3(s)} + 3\,H_{2(g)}$$

The masses of each species in the above reaction to react exactly are given by

$$m(\mathrm{Al}) = 2 \times 26.981\,539\ \mathrm{g}$$
$$m(\mathrm{HCl}) = 6 \times (1.007\,94 + 35.4527)\ \mathrm{g}$$

(a) Evaluate $m(\mathrm{Al})$ and $m(\mathrm{HCl})$.
(b) If 300 g of HCl reacts with $m(\mathrm{Al})$, what mass of HCl remains unreacted?
(c) If 100 g of Al reacts with $m(\mathrm{HCl})$, what mass of Al remains unreacted?

Give all answers to the appropriate level of precision.

6. Uncertainties

We have seen in earlier chapters that the same quantity can be quoted to various degrees of precision. The uncertainty in a number is however a measure of its accuracy, i.e. how close it is to the true value. Determination of uncertainties is important when quantities are measured in the laboratory.

Suppose we measure a true value of 4.87 as 4.96. The absolute error is then the difference between the measured value and the true value, in this case $4.96 - 4.87$ or 0.09. Note that if the measured value is less than the true value the absolute error is negative, e.g. if the true value of 4.87 is measured as 4.52 the absolute error is $4.52 - 4.87$ or -0.35.

The fractional error is simply the absolute error divided by the actual value. In the first example, this is 0.09/4.87 or 0.018, in the second, $-0.35/4.87$ or -0.072. The fractional error is converted to a percentage error by multiplying by 100, so the respective percentage errors for these examples are 1.8% and −7.2%.

In most real cases of measurement we don't know the true value, so we need to estimate the uncertainty of measurement from the instrument being used. This gives an absolute error, which is divided by the measured value to give the fractional and percentage errors as above.

Note that changing from the true to the measured value makes very little difference to the fractional and percentage errors. In the first case above we would have a fractional error of 0.09/4.96 or 0.018 and a percentage error of 1.8%, in the second a fractional error of -0.077 and a percentage error of -7.7%.

Systematic errors

The treatment above assumes that the uncertainty of a measurement has an equal chance of being positive or negative. If, however, an instrument has a zero error, then a constant correction has to be applied to each measurement before we can consider the effect of these random uncertainties. For example, if we know that an instrument reads 0.2 when it should read 0.0, we first need to subtract 0.2 from each reading to give the true value.

Examples from the chemistry laboratory

(a) The markings on a burette are typically 0.10 cm^3 apart. It should therefore be possible to read the burette to the nearest 0.05 cm^3. Because we don't know whether our readings will be higher or lower than the true uncertainty, we denote this uncertainty as $\pm 0.05\ cm^3$. This will be the absolute error for any reading of this burette.

If a burette reading is given as 27.35 cm^3, the fractional error will be $\pm 0.05\ cm^3/27.35\ cm^3$ or 0.0018 and the percentage error 100 times this or 0.18%. Note that the fractional and percentage errors never have units since they are calculated by taking the ratio of two quantities having the same units, i.e. the units cancel.

(b) When accurate values of mass are required we would use a balance reading to the nearest 0.0001 g. The uncertainty in such readings is then probably of the order of half a division, or $\pm 0.000\,05$ g. Note again the use of the plus or minus symbol ($\pm$) as we don't know whether we are reading values too high or too low. The absolute error on such readings is therefore $\pm 0.000\,05$ g. Let us consider the fractional and percentage errors on some typical masses that would be measured using such an instrument as shown in the table overleaf.

This table shows two things. First, this is a very precise instrument as for reasonable masses the apparent error is well below 1%. Second, not surprisingly, the relative errors become less as the magnitude of the quantity being measured increases. Thus it is always desirable to measure as large a quantity as possible if circumstances allow.

Suppose now that the balance has a zero error, and reads 0.0007 g when it should in fact read 0.0000 g. A value that we read as 2.4218 g then needs to be corrected by subtracting the zero error, i.e. by calculating 2.4218 g – 0.0007 g to give 2.4211 g. Note that zero errors can be positive or negative, but are always constant for a given individual instrument.

Mass/g	Fractional error	Percentage error, %
0.0100	0.005	0.5
0.100	0.0005	0.05
1.00	0.000 05	0.005
10.0	0.000 005	0.0005

Questions

1. A true value of 16.87 is measured as 16.72. Calculate:
   (a) the absolute error
   (b) the fractional error
   (c) the percentage error.

2. A true wavelength of 472 nm is measured as 482 nm. Calculate:
   (a) the absolute error
   (b) the fractional error
   (c) the percentage error.

3. A voltmeter measures the EMF of an electrochemical cell to the nearest 0.1 V. Assuming an uncertainty of ± 0.05 V on a reading of 6.45 V, calculate:
   (a) the absolute error
   (b) the fractional error
   (c) the percentage error.

4. A polarimeter has a zero error of -0.1°, i.e. it reads -0.1° when the true angle of rotation is zero. What is the true angle when the polarimeter reads:
   (a) 2.3°
   (b) 4.6°
   (c) 9.8°

5. In a kinetics experiment a clock reads to the nearest second, so the uncertainty can be taken as ± 0.5 s. In addition it shows $+0.5$ s when the true value is zero. When the clock shows 37.0 s,
   (a) what is the true value of the time
   (b) what is the absolute error
   (c) what is the fractional error
   (d) what is the percentage error?

7. Maximum Possible Error

Although it is relatively straightforward to estimate the uncertainty on an individual measurement, many of the quantities we wish to calculate involve the combination of these individual measurements in some way. The question then arises as to how the uncertainties combine when the quantities to which they refer are combined.

The maximum possible error allows us to track the propagation of a set of uncertainties by considering the worse case scenario each time. Consider the following examples, where we are combining two quantities measured as 12.3 ± 0.2 and 3.7 ± 0.4. Thus the first quantity varies between 12.1 and 12.5, and the second between 3.3 and 4.1.

Summation

If these quantities are added, the maximum value of the sum is found by taking the maximum value of each of the measured quantities, i.e. $12.5 + 4.1$ or 16.6. The minimum value of the sum is found by taking the minimum value of each of the measured quantities, i.e. $12.1 + 3.3$ or 15.4. The calculated quantity is quoted as 16.0 ± 0.6, the quoted error being half the difference between the maximum and minimum values.

Difference

If we now take the difference $12.3 - 3.7$, careful thought shows that the maximum value of the difference is found by taking $3.7 - 0.4$ from $12.3 + 0.2$, i.e. using the two extreme values. This gives $12.5 - 3.3$ or 9.2. The minimum value of the difference is found by subtracting $3.7 + 0.4$ from $12.3 - 0.2$, i.e. the difference between the closest values or 8.0. Overall then the difference is quoted as being 8.6 ± 0.6.

Product

When the two quantities are multiplied, we again find the maximum value by multiplying the individual maxima, and the minimum value by multiplying the individual minima. Consequently the maximum value is 12.5×4.1 or 51, while the minimum is 12.1×3.3 or 40. Hence the product is quoted as 46 ± 5.

Quotient

To determine the uncertainty in the quotient 12.3/3.7, careful thought again tells us that the largest value is obtained by dividing the maximum value of 12.5 by the minimum value of 3.3 to give 3.8. Similarly, the smallest value is obtained by dividing the minimum value of 12.1 by the maximum value of 4.1 to give 3.0. The quotient is now quoted as 3.4 ± 0.5.

Calculation of overall pressure

In a mixture of two gases, the partial pressure of hydrogen is measured as 1.78 ± 0.06 atm, while that of nitrogen is measured as 2.42 ± 0.08 atm. The overall pressure of the mixture is given by the sum of the partial pressures. The maximum value of this is found by adding the two maxima, i.e. $1.84 + 2.50$ or 4.34 atm, while the minimum value is found by adding the two minima, i.e. $1.72 + 2.34$ or 4.06 atm. The overall pressure is quoted as 4.20 ± 0.14 atm.

Calculating a titre

In a titration, the initial burette reading was 0.45 ± 0.05 cm^3 and the final burette reading 22.60 ± 0.05 cm^3. The maximum value of the titre is given by calculating $22.65 - 0.40$ cm^3, which gives 22.25 cm^3. The minimum value is given by $22.55 - 0.50$ cm^3 or 22.05 cm^3. The overall titre is thus quoted as 22.15 ± 0.10 cm^3.

If the titre contains hydrochloric acid of concentration 0.103 ± 0.002 mol dm^{-3}, we can calculate the amount of acid delivered by multiplying the volume by the concentration. The maximum amount of acid is therefore 22.25 cm^3 $\times$ 0.105 mol dm^{-3} or 0.00234 mol, and the minimum 22.05 cm^3 $\times$ 0.101 mol dm^{-3} or 0.002 23 mol. The amount of acid delivered is therefore recorded as $0.00229 \pm 0.000\,06$ mol.

Calculation of entropy

The entropy ΔS of a phase change is found by dividing the corresponding enthalpy change ΔH by the temperature T at which the change takes place. The enthalpy of fusion of silver is 11.30 ± 0.05 kJ mol^{-1} at a temperature of 1234 ± 10 K. The maximum value for the entropy is thus given by 11.35×10^3 J mol^{-1}/1224 K or 9.273 J K^{-1} mol^{-1}, while the minimum value is found from 11.25×10^3 J mol^{-1}/1244 K or 9.043 J K^{-1} mol^{-1}. The entropy change could thus be quoted as 9.158 ± 0.115 J K^{-1} mol^{-1}. However, it makes more sense to give it as 9.2 ± 0.1 J K^{-1} mol^{-1}, as there is little point in quoting uncertainties to a high level of precision.

Questions

1. For carbon dioxide the enthalpy of fusion $\Delta_{fus}H$ is 8.3 ± 0.1 kJ mol^{-1}, while the enthalpy of vaporisation $\Delta_{vap}H$ is 16.9 ± 0.2 kJ mol^{-1}. In fact, carbon dioxide sublimes with the enthalpy of sublimation $\Delta_{sub}H$ given by

$$\Delta_{sub}H = \Delta_{fus}H + \Delta_{vap}H$$

What is the maximum possible error in $\Delta_{sub}H$ for carbon dioxide?

2. The half cell reaction

$$Zn_{(s)} \rightarrow Zn^{2+}{}_{(aq)} + 2e^-$$

has an electrode potential $E^\ominus$ of 0.76 ± 0.01 V, while $E^\ominus$ for the reaction

$$Cu^{2+}{}_{(aq)} + 2e^- \rightarrow Cu_{(s)}$$

is 0.34 ± 0.01 V. The overall EMF for the cell reaction

$$Cu^{2+}{}_{(aq)} + Zn_{(s)} \rightarrow Cu_{(s)} + Zn^{2+}$$

is found by adding the two $E^\ominus$ values above. What is the maximum possible uncertainty in the overall EMF?

3. Lead(II) chloride is slightly soluble in water due to the equilibrium

$$PbCl_{2(s)} \rightleftharpoons Pb^{2+}{}_{(aq)} + 2Cl^-{}_{(aq)}$$

The solubility product K_s is equal to $4s^3$, where s is the solubility of $PbCl_2$. If $s = (1.62 \pm 0.02) \times 10^{-2}$ mol dm^{-3}, what is the maximum possible error in K_s?

4. The density ρ of a substance can be found from the formula

$$\rho = \frac{m}{V}$$

where m represents the mass and V the volume. What is the density of gold, if a sample of volume 27 ± 1 cm^3 has a mass of 521 ± 5 g?

5. For the reaction

$$2NO_{2(g)} \rightleftharpoons N_2O_{4(g)}$$

$\Delta H^\ominus = -58.0 \pm 0.5$ kJ mol^{-1} and $\Delta S^\ominus = -177 \pm 1$ J K^{-1} mol^{-1} at a temperature of 298 ± 1 K. If $\Delta G^\ominus = \Delta H^\ominus - T\Delta S^\ominus$, what is the maximum possible error in $\Delta G^\ominus$?

8. Maximum Probable Error

In Chapter 7 we saw that the maximum possible error allows us to estimate the worst case scenario. The maximum probable error, however, takes into account the fact that this rarely is the case, with the error usually being somewhat smaller. It is straightforward to calculate the maximum probable error using the following formulae. In each case we assume that the quantities X and Y are measured, and are used to calculate Z. The estimated absolute errors on each quantity are ΔX, ΔY and ΔZ respectively.

Sums and differences

If $Z = X + Y$ or $Z = X - Y$ or $Z = Y - X$, the absolute error ΔZ on Z is given by

$$\Delta Z = \surd[(\Delta X)^2 + (\Delta Y)^2]$$

Using the example from Chapter 7, where $X = 12.3 \pm 0.2$ and $Y = 3.7 \pm 0.4$, the error ΔZ will be given by

$$\begin{aligned}\Delta Z &= \surd(0.2^2 + 0.4^2)\\ &= \surd(0.04 + 0.16)\\ &= \surd 0.20\\ &= 0.45\end{aligned}$$

Using an appropriate number of figures, the difference is then quoted as 8.6 ± 0.5. Note that this is slightly less than the maximum possible error calculated previously.

Products and quotients

The rule here is similar to that above, but in this case the fractional errors are combined rather than the absolute errors. The fractional error $\Delta Z/Z$ on Z is given by

$$\frac{\Delta Z}{Z} = \sqrt{\left[\left(\frac{\Delta X}{X}\right)^2 + \left(\frac{\Delta Y}{Y}\right)^2\right]}$$

Again using the example of a product from Chapter 7, where 12.3 ± 0.2 (X) is multiplied by 3.7 ± 0.4 (Y), the fractional error $\Delta Z/Z$ is given by

$$\begin{aligned}\frac{\Delta Z}{45.51} &= \sqrt{\left[\left(\frac{0.2}{12.3}\right)^2 + \left(\frac{0.4}{3.7}\right)^2\right]}\\ &= \surd(0.0162^2 + 0.1081^2)\\ &= \surd(0.000\,262\,44 + 0.011\,69)\\ &= \surd(0.011\,948)\\ &= 0.1093\end{aligned}$$

At this point it is important to realise that we have calculated the fractional error. The absolute error ΔZ is found by rearranging this equation to give

$$\Delta Z = 45.51 \times 0.1093 = 4.97$$

Using an appropriate number of figures the final value can be quoted as 46 ± 5.

Calculation of overall pressure

Using the example involving partial pressures from Chapter 7, which involves summation, the absolute error will be given by

$$\sqrt{[(0.06\,\text{atm})^2 + (0.08\,\text{atm})^2]} = \sqrt{(0.0036\,\text{atm}^2 + 0.0064\,\text{atm}^2)}$$
$$= \sqrt{(0.0100\,\text{atm}^2)}$$
$$= 0.1\,\text{atm}$$

The overall pressure is given as 4.20 ± 0.10 atm. This is less than the maximum possible error for this calculation.

Calculation of entropy

Again using an example from Chapter 7, which involves calculation of a quotient, the fractional error will be given by

$$\frac{\Delta(\Delta S)}{9.158\,\text{J K}^{-1}\,\text{mol}^{-1}} = \sqrt{\left[\left(\frac{0.05\,\text{kJ mol}^{-1}}{11.30\,\text{kJ mol}^{-1}}\right)^2 + \left(\frac{10\,\text{K}}{1234\,\text{K}}\right)^2\right]}$$
$$= \sqrt{(0.004\,425^2 + 0.008\,103^2)}$$
$$= \sqrt{(0.000\,019\,58 + 0.000\,065\,66)}$$
$$= \sqrt{(0.000\,085\,24)}$$
$$= 0.009\,232$$

Rearranging this gives

$\Delta(\Delta S) = 9.158\ \text{J K}^{-1}\ \text{mol}^{-1} \times 0.009\,232 = 0.0846\ \text{J K}^{-1}\ \text{mol}^{-1}$

The final answer can be quoted as $\Delta S = 9.16 \pm 0.08\ \text{J K}^{-1}\ \text{mol}^{-1}$.

Questions

1. When 1 mol of ice at 0°C is converted to steam at 100°C, the entropy change ΔS has to be calculated in three stages:

   (i) ΔS_1 is the entropy change for the melting of ice at 0°C,
   (ii) ΔS_2 is the entropy change for the temperature increase from 0°C to 100°C,
   (iii) ΔS_3 is the entropy change for the vaporisation of water at 100°C.

   These are combined so that overall

   $$\Delta S = \Delta S_1 + \Delta S_2 + \Delta S_3$$

   If $\Delta S_1 = 22.007 \pm 0.004\ \text{J K}^{-1}\ \text{mol}^{-1}$, $\Delta S_2 = 98.38 \pm 0.1\ \text{J K}^{-1}\ \text{mol}^{-1}$ and $\Delta S_3 = 109.03 \pm 0.2\ \text{J K}^{-1}\ \text{mol}^{-1}$, determine the maximum probable error in ΔS.

2. In a particular hydrogenlike atom, spectral lines appear with wavelengths 19.440 ± 0.007 nm and 17.358 ± 0.005 nm. What is the maximum probable error in the difference between these wavelengths?

3. The rate constant k is $(9.3 \pm 0.1) \times 10^{-5}\text{s}^{-1}$ for the reaction

   $$NH_2NO_{2(aq)} \rightarrow N_2O_{(g)} + H_2O_{(l)}$$

   The rate of this reaction depends on the concentration of NH_2NO_2 according to the equation

   $$Rate = k[NH_2NO_2]$$

   Determine the maximum probable error in the rate if $[NH_2NO_2] = 0.105 \pm 0.003\ \text{mol dm}^{-3}$.

4. The entropy of vaporisation $\Delta_{\text{vap}}S$ is given by the equation

   $$\Delta_{\text{vap}}S = \frac{\Delta_{\text{vap}}H}{T_b}$$

where $\Delta_{vap}H$ is the enthalpy of vaporisation and T_b the boiling temperature. Determine the maximum probable error in $\Delta_{vap}S$ for $CHCl_3$ for which $\Delta_{vap}H = 29.4 \pm 0.1$ kJ mol^{-1} and $T_b = 334 \pm 1$ K.

5. The enthalpy change ΔH for a gaseous reaction can be given in terms of the change in the amount of gaseous species Δn as

$$\Delta H = \Delta U + RT\,\Delta n$$

where ΔU is the change in internal energy, R the gas constant and T the absolute temperature. For the reaction

$$C_2H_5OH_{(l)} + 3O_{2(g)} \rightarrow 2CO_{2(g)} + 3H_2O_{(l)}$$

$\Delta n = -1$ mol, which is an exact value. If $\Delta U = -1364.2 \pm 0.1$ kJ mol^{-1} and $T = 298 \pm 1$ K, determine the maximum probable error in ΔH. Take R as 8.31 ± 0.05 J K^{-1} mol^{-1}.

9. Simple Statistics

In the previous three chapters we have considered how to estimate uncertainties on a given quantity starting with an estimate of the uncertainties in the original measurement. We now turn our attention to how this can be done if we have several repeat measurements of the same quantity. To do this we need to apply some simple statistical techniques.

Consider the following two sets of data:

21 24 25 26 29 and 4 9 23 41 48

Both have an average of 25, found by adding up the five numbers in the set and dividing by 5. However, if these were a set of repeat measurements of a particular quantity then we would probably consider the first set to be more reliable, as they are all closer to the average than those in the second set. They also have a smaller spread, as given by the difference between the maximum and minimum values.

Standard deviation

This spread can be quantified by calculating the standard deviation, s, for each set of data. This can be found from the equation:

$$s^2 = \sum \frac{(x_i - \bar{x})^2}{n-1}$$

This equation may seem quite complicated at first, but it is easier to apply once we understand the meaning of all the symbols. First of all, x_i stands for the individual measurements made. There are 5 values of x, so i takes the values 1, 2, 3, 4, and 5. We thus have, in the first set of data, $x_1 = 21$, $x_2 = 24$, $x_3 = 25$, $x_4 = 26$, and $x_5 = 29$. The quantity $\bar{x}$ is the average of these, which we have already established as 25. The number of data is n, in this case 5. The symbol Σ (capital Greek sigma) represents a summation, so we simply add all the terms which appear to the right of it. It is easiest to do this using a table, as shown below.

i	x_i	$x_i - \bar{x}$	$(x_i - \bar{x})^2$
1	21	−4	16
2	24	−1	1
3	25	0	0
4	26	1	1
5	29	4	16

The summation sign in the equation tells us to add up all the terms in the far right column, which gives 34. Note that the individual terms are all positive since the square of a negative number is positive.

Substituting into the equation now gives

$$s^2 = \frac{34}{5-1} = \frac{34}{4} = 8.5$$

Taking the square root of this gives $s = 2.9$.

How does this value relate to the spread of the data? To answer this we need to consider the behaviour of large amounts of data.

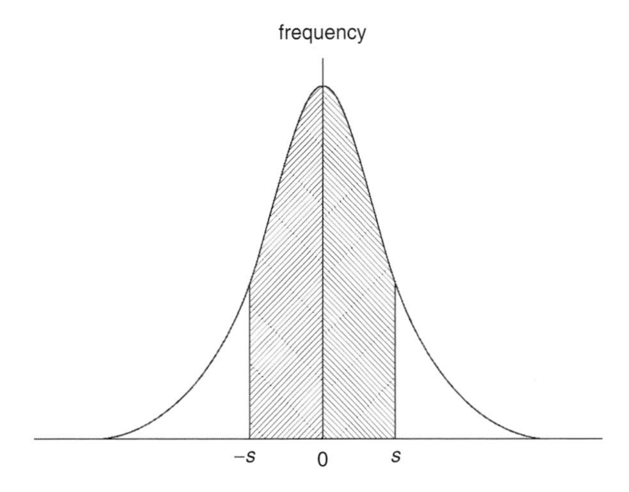

Figure 9.1 Normal distribution showing frequency of occurrences of a particular value against that value. The shaded area indicates those values which fall within one standard deviation of the mean

Normal distribution

The normal distribution is illustrated in Fig. 9.1, which shows the number of occurrences against its value x_i. Note that the greatest number of occurrences would be expected for the average $\bar{x}$. The x-axis of the graph is marked in numbers of standard deviations. The probability of a measurement falling within one standard deviation of the average is given by the area shaded in Fig. 9.1; this is 68% of the total. More usually twice the standard deviation would be quoted as a reasonable uncertainty limit, as there is a 95% probability of our value of x_i being within this, as shown in Fig. 9.2.

The carbon–hydrogen bond

The bond dissociation enthalpies ΔH of the carbon–hydrogen bond in a series of environments is as follows:

Environment	ΔH/kJ mol^{-1}
CH_4	438
$-CH_3$	465
$-CH_2-$	422
$=CH-$	339
C_2H_6	420

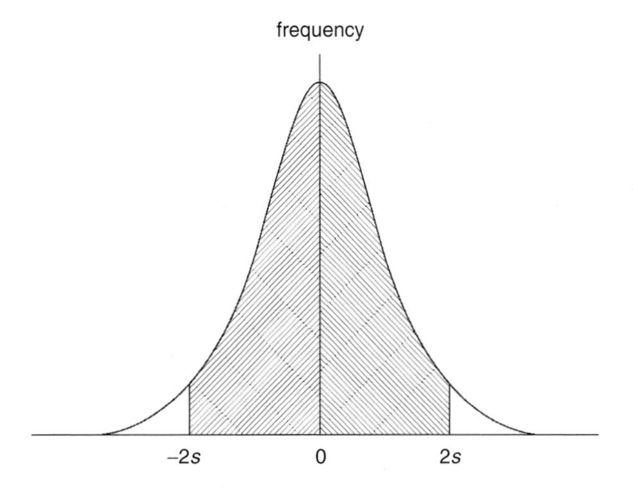

Figure 9.2 The same graph as in Fig. 9.1, with the shaded area now showing those values which fall within two standard deviations of the mean

The first step in determining the standard deviation of these values is to calculate the average, given by

$$\frac{438 + 465 + 422 + 339 + 420}{5} \quad \text{or} \quad 417$$

We then draw up a table as before:

i	ΔH_i/kJ mol^{-1}	$(\Delta H_i - \Delta H)$/kJ mol^{-1}	$(\Delta H_i - \Delta H)^2$/kJ^2 mol^{-2}
1	438	21	441
2	465	48	2304
3	422	5	25
4	339	−78	6084
5	420	3	9

In this case the sum of the $(\Delta H_i - \Delta H)^2$ in the right-hand column is 8863 kJ^2 mol^{-2}, allowing us to calculate s^2 as

$$s^2 = \frac{8863\,\text{kJ}^2\,\text{mol}^{-2}}{5-1}$$
$$= 2216\,\text{kJ}^2\,\text{mol}^{-2}$$

and hence $s = 47$ kJ mol^{-1}. It would thus be reasonable to quote ΔH for this set of environments as $\Delta H = 417 \pm 94$ kJ, mol^{-1}, using two standard deviations as a reasonable estimate of the uncertainty, as explained previously.

Questions

1. The concentration of lead (in μg m^{-3}) was measured at a site on consecutive days to have the values:

   0.380 0.311 0.305 0.233 0.335 0.370

   Calculate the standard deviation without using a calculator.

2. The concentration of magnesium (in mg dm^{-3}) in a series of samples of surface water had the values:

   0.837 0.409 0.621 1.357 0.723 0.834 1.041 1.454

   Calculate the standard deviation without using a calculator.

3. A series of measurements of atmospheric pressure (in kPa) gave the values:

   1014.46 1015.34 1015.66 1014.50 1015.89

   Calculate the standard deviation without using a calculator.

4. A series of Cu–O bond lengths (in Å) were determined by X-ray crystallography in a compound and had the values:

   1.982 1.969 2.204 1.935 1.999 2.186 2.067

   Calculate the standard deviation without using a calculator.

10. The *t*-Statistic

The use of the standard deviation to determine an uncertainty limit is based on the assumption that we have a reasonably large sample, i.e. greater than 25. If that is not the case we run into the problem that the average we have determined is unlikely to be close enough to the true average, which is our required value. This is overcome by calculating the quantity

$$\frac{ts}{\sqrt{n}}$$

In this expression, as before, s is the calculated standard deviation and n is the sample size. The quantity t is known as the t-value, and can be determined from statistical tables. Its precise value depends on the level of accuracy required together with a quantity known as the number of degrees of freedom, which is equal to $n - 1$. Relevant values of t are given in Appendix 5.

If we perform this calculation for the first data set given in Chapter 9, we have a confidence interval of

$$\frac{2.78 \times 2.9}{\sqrt{5}} = \frac{2.78 \times 2.9}{2.24} = 3.6$$

since the value of t for a 95% confidence level and 4 degrees of freedom is 2.78 and so we quote the average of the data as being 25 ± 4.

The carbon-hydrogen bond

Applying this method to the data used in the previous chapter, which also has $t = 2.78$, gives a confidence interval of

$$\frac{2.78 \times 47\,\text{kJ mol}^{-1}}{\sqrt{5}} = \frac{2.78 \times 47\,\text{kJ mol}^{-1}}{2.24} = 58\,\text{kJ mol}^{-1}$$

and so the final value quoted is $\Delta H = 417 \pm 58$ kJ mol^{-1}.

Wavelengths of absorption in dye molecules

The series of dyes represented by the formula in Fig. 10.1 exhibit different colours depending on the nature of the substituent X. These can be quantified in terms of the wavelength λ_{max} at which most light is absorbed, as shown in the following table.

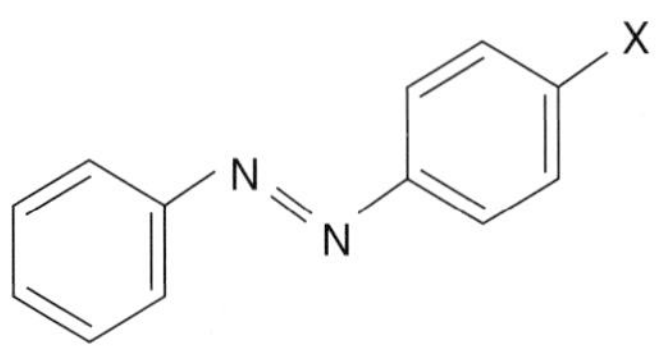

Figure 10.1 Structure of a dye molecule with variable substituent X

X	λ_{max}/nm
H	318
CH_3	333
NO_2	338
$(C_2H_5)_2N$	415
CH_3O	346
NC	324
NH_2	387
SO_3H	329
Br	329
COOH	325

In order to determine the uncertainty limits on the average value of λ_{max}, we need to calculate the standard deviation.

We start by determining the average, which is 344 nm. As before, we now draw up a table with all the required quantities.

X	λ_{max}	$(\lambda_{max} - \bar{\lambda}_{max})$/nm	$(\lambda_{max} - \bar{\lambda}_{max})^2$/nm^2
H	318	−26	676
CH_3	333	−11	121
NO_2	338	−6	36
$(C_2H_5)_2N$	415	71	5041
CH_3O	346	2	4
NC	324	−20	400
NH_2	387	43	1849
SO_3H	329	15	225
Br	329	15	225
COOH	325	−19	361

The sum of the $(\lambda_{max} - \bar{\lambda}_{max})^2$ values, shown in the far right column, is 8938 nm^2. Thus

$$s^2 = \frac{8938}{10-1} = \frac{8938}{9} = 993\,\text{nm}^2$$

and so $s = 31.5$ nm.

At this point we need to realise that we have too few readings for the standard deviation to be a satisfactory indication of the uncertainty, and that we need to use the t-statistic to generate this. The appropriate value of t for a 95% confidence level with 10 − 1 or 9 degrees of freedom is 2.26. The confidence limit is therefore

$$\frac{2.26 \times 31.5\,\text{nm}}{\sqrt{10}} = \frac{71.19\,\text{nm}}{3.16} = 22.5\,\text{nm}$$

and the average value of λ_{max} is 344 ±23 nm.

Questions

1. In a quantum mechanical computer simulation of a series of alkylamines, the following values (in Å) were obtained for the C–N bond length:

   1.456 1.454 1.459 1.465 1.512 1.522 1.517

   Determine the standard deviation using a calculator, and hence calculate the 90% confidence limit for these data.

2. The analysis of NO_2 in the atmosphere gave the following successive readings (in parts per billion):

   3.62 5.15 4.26 5.42 3.48

   Determine the standard deviation using a calculator, and hence calculate the 95% confidence limit for these data.

3. The measurement of the pH of rainwater at a certain site gave the following data for different samples:

   5.53 5.32 4.92 5.16 4.93 5.13 4.90 5.06

   Determine the standard deviation using a calculator, and hence calculate the 95% confidence limit for these data.

4. The concentration of aluminium in rainwater (in mg dm^{-3}) was measured in successive samples as:

   0.017 0.026 0.005 0.013 0.026 0.019

   Determine the standard deviation using a calculator, and hence calculate the 97.5% confidence limit for these data.

5. Successive readings of atmospheric pressure gave the following values (in kPa):

   1030.45 1029.56 1029.17 1029.01 1029.50

   Determine the standard deviation using a calculator, and hence calculate the 99% confidence limit for these data.

Algebra

11. Precedence

In an expression such as

$$2 + 8 - 4$$

we would probably add 2 and 8 to give 10, then subtract 4 to give 6. We could also subtract 4 from 8 to give 4, and add this to 2 to give 6. We could even subtract 4 from 2 to give -2, then add 8 to give 6. Clearly the order in which we perform these operations doesn't matter in this case.

Suppose that the expression is now

$$2 \times 8 - 4$$

We can now multiply 2 and 8 to give 16, then subtract 4 to give 12. On the other hand, we could subtract 4 from 8 to give 4, then multiply by 2 to give 8. Clearly in this case the order in which we perform the operations does matter. Which way is correct?

BODMAS rules

The BODMAS acronym stands for Brackets, Of, Division, Multiplication, Addition, Subtraction and gives us the order in which the operations should be performed in a given expression.

- *Brackets* can be used to alter the order of the operations. Any expression within brackets is evaluated first.
- *Of* is typically used with fractions, as in 1/2 of 6. In terms of operations it is equivalent to multiply.
- *Multiplication* and *division* have equal precedence. Note that raising a number to a power is equivalent to multiplication, as for example in $2^3 = 2 \times 2 \times 2$.
- *Addition* and *subtraction* have equal precedence.

These rules show that $2 \times 8 - 4$ is correctly evaluated as 12, as 2 is multiplied by 8 first of all and 4 is then subtracted.

Examples

(i) $2 + 3 \times 6 - 4$

The first operation to perform is multiplication, i.e. $3 \times 6 = 18$. Then add 2 to give 20 and subtract 4 to give 16.

(ii) $8 - 9/3^2$

In this case, evaluate 3^2, which is 3×3 or 9. The expression then becomes $8 - 9/9$. Perform the division $9/9 = 1$, and so we are left with $8 - 1$ or 7.

Density

The density ρ of a solution of methanol in water having a mole fraction x of methanol can be approximated as

$$\rho/\mathrm{g\,cm^{-3}} = 0.9971 - 0.289\,30x + 0.299\,07x^2$$

To decide on the order of precedence in which this expression is evaluated, first write x^2 as $x \times x$ to give

$$\rho/\mathrm{g\,cm^{-3}} = 0.9971 - 0.289\,30x + 0.299\,07x \times x$$

If we now realise that $0.289\,30x$ actually means $0.289\,30 \times x$, i.e. some of the multiplication signs are omitted, we can write

$$\rho/\mathrm{g\,cm^{-3}} = 0.9971 - 0.289\,30 \times x + 0.299\,07 \times x \times x$$

If we wish to know the value of ρ when $x = 0.25$, we can substitute to give

$$\rho/\mathrm{g\,cm^{-3}} = 0.9971 - 0.289\,30 \times 0.25 + 0.299\,07 \times 0.25 \times 0.25$$

The BODMAS rules indicate that multiplication is performed before addition and subtraction, so evaluating gives

$$0.289\,30 \times 0.25 = 0.072\,33 \quad \text{and} \quad 0.299\,07 \times 0.25 \times 0.25 = 0.018\,69$$

Substituting gives

$$\rho/\mathrm{g\,cm^{-3}} = 0.9971 - 0.072\,33 + 0.018\,69 = 0.943\,46$$

so when $x = 0.25$, the value of ρ is 0.943 46 g cm^{-3}. However, this value needs to be rounded to 2 significant figures and then 3 decimal places consistent with the value of x, to give ρ as 0.94 g cm^{-3}.

Two-dimensional harmonic oscillator

The potential energy, V, of a two-dimensional harmonic oscillator is given by the equation:

$$V = \frac{k}{2}(x^2 + y^2)$$

where x and y denote the position defined in two perpendicular directions. The constant k is known as the force constant and is a measure of the flexibility of the oscillator.

Suppose that we wish to determine the potential energy when $x = 1.5$ nm and $y = 0.5$ nm for a two-dimensional harmonic oscillator in which $k = 250$ N m^{-1}. The equation can be rewritten to show the implied multiplication as

$$V = \frac{k}{2} \times (x \times x + y \times y)$$

Remember that the BODMAS rules indicate that the expression in brackets is evaluated first. This is $x \times x + y \times y$, where the multiplications are performed first. Hence

$$x \times x = 1.5\,\mathrm{nm} \times 1.5\,\mathrm{nm} = 2.25\,\mathrm{nm^2}$$

and

$$y \times y = 0.5\,\mathrm{nm} \times 0.5\,\mathrm{nm} = 0.25\,\mathrm{nm^2}$$

The quantity within the brackets is then

$$2.25\,\mathrm{nm^2} + 0.25\,\mathrm{nm^2} = 2.50\,\mathrm{nm^2}$$

In the rest of the expression k is divided by 2 to give

$$250\,\mathrm{N\,m^{-1}}/2 = 125\,\mathrm{N\,m^{-1}}$$

Then multiply these quantities together

$$125\,\mathrm{N\,m^{-1}} \times 2.50\,\mathrm{nm^2}$$

Since 1 nm = 10^{-9} m, it follows that 1 nm^2 = 1 nm $\times$ 1 nm = 10^{-9} m $\times$ 10^{-9} m = 10^{-18} m^2, so the final value is

$$125\,\mathrm{N\,m^{-1}} \times 2.50 \times 10^{-18}\,\mathrm{m^2}$$

which is equal to 3.13×10^{-16} N m or 3.13×10^{-16} J, since 1 N m = 1 J (see Appendix 2).

Hydrogenation of ethene on copper

The rate of a chemical reaction is measured as the rate at which the concentration of a reactant decreases, with respect to time. Its units are mol dm^{-3} s^{-1} for any reaction. The rate is generally related to the concentrations of reactants, usually denoted by the use of square brackets, by one or more rate constants. The rate of hydrogenation of ethene on copper is governed by the rate equation

$$Rate = \frac{k_a[H_2][C_2H_4]}{(1 + k_b[C_2H_4])^2}$$

where k_a and k_b are rate constants, $[H_2]$ is the concentration of hydrogen and $[C_2H_4]$ is the concentration of ethene.

Consider the calculation of the rate when $[H_2] = 0.05\ \text{mol dm}^{-3}$ and $[C_2H_4] = 0.15\ \text{mol dm}^{-3}$, given the constants

$$k_a = 4 \times 10^5\ \text{dm}^3\ \text{mol}^{-1}\ \text{s}^{-1}$$
$$k_b = 1.62\ \text{dm}^3\ \text{mol}^{-1}$$

First of all rewrite the equation to include the implied multiplication operations:

$$Rate = \frac{k_a \times [H_2] \times [C_2H_4]}{(1 + k_b \times [C_2H_4])^2}$$

Evaluate the bracket first, which contains addition and multiplication operations. Perform the multiplication first to give

$$k_b \times [C_2H_4] = 1.62\ \text{dm}^3\ \text{mol}^{-1} \times 0.15\ \text{mol dm}^{-3} = 0.243$$

noting that the units cancel in this case. The second operation in the bracket is the addition, which gives

$$1 + 0.243 = 1.243$$

The remaining operations are multiplications and divisions, so we can substitute in the original equation to give

$$Rate = \frac{4 \times 10^5\ \text{dm}^3\ \text{mol}^{-1}\ \text{s}^{-1} \times 0.05\ \text{mol dm}^{-3} \times 0.15\ \text{mol dm}^{-3}}{1.243^2}$$

The numerical part of the top of this fraction is evaluated as 3×10^3, while the units can be shown to combine to give $\text{mol dm}^{-3}\ \text{s}^{-1}$. The final calculation required is thus

$$Rate = \frac{3 \times 10^3\ \text{mol dm}^{-3}\ \text{s}^{-1}}{1.545} = 1.9 \times 10^3\ \text{mol dm}^{-3}\ \text{s}^{-1}$$

Questions

1. The equilibrium constant K for a certain reaction is given by

$$K = \frac{16x^2(1-x)}{(1-3x)^3\left(\frac{p}{p^{\circ}}\right)^2}$$

where x is the extent of reaction, p is the pressure, and $p^{\circ} = 1$ atm. Calculate the value of K when $x = 0.15$ and $p = 2.5$ atm.

2. When determining the variation of equilibrium constant K with temperature T, it is necessary to evaluate the quantity

$$\frac{\Delta H^{\circ}}{R}\left(\frac{1}{T_1} - \frac{1}{T_2}\right)$$

Evaluate this expression when $\Delta H^{\circ} = 38.4\ \text{kJ mol}^{-1}$, $T_1 = 298$ K and $T_2 = 300$ K. (The value of the gas constant R can be found in Appendix 4.

3. The equation for the dew point line is

$$p = \frac{p_1^* p_2^*}{p_1^* + (p_2^* - p_1^*)y_1}$$

where the overall pressure p is given in terms of the vapour pressures p_1^* and p_2^* of components 1 and 2 respectively, and y_1 is the mole fraction of component 1 in the vapour. Calculate p, if $p_1^* = 3125$ Pa, $p_2^* = 2967$ Pa and $y_1 = 0.365$.

4. The volume V of a solution of glycerol in water is given by the equation

$$V/\text{cm}^3 = 18.023 + 53.57x + 1.45x^2$$

where x is the mole fraction of glycerol. Calculate V, if $x = 0.27$.

12. Fractions

The meaning of a fraction such as a/b is simply 'a divided by b'. Many fractions therefore have exact decimal equivalents, such as $^1/_4 = 0.25$ or $1/10 = 0.1$. Others do not, such as $1/3$ and $22/7$, and in many cases we choose to leave these in fractional form to retain their exact values.

The correct terminology to use when describing a fraction is 'numerator over denominator'. In a/b, the numerator is a while the denominator is b. It is useful to note that if both numerator and denominator are multiplied by a common factor the fraction remains unchanged. For example,

$$\frac{1}{2} = \frac{1 \times 2}{2 \times 2} = \frac{2}{4} = 0.5$$

Both numerator and denominator can also be divided by a common factor to leave the fraction unchanged.

The reciprocal of a fraction simply changes the position of the numerator and the denominator. Thus the reciprocal of $2/3$ is $3/2$, and that of x/y is y/x. Note that since 9 can be thought of as $9/1$, its reciprocal is $1/9$.

Combinations of fractions

It is important to be able to combine fractions, particularly when they are expressed in symbols or when there is no exact decimal equivalent.

Addition and subtraction

The important thing to remember with addition and subtraction is that we can only calculate the sum or difference of a pair of fractions if they are expressed in terms of a common denominator. Thus if we wish to calculate

$$\frac{1}{2} + \frac{2}{3}$$

we need to realise that

$$\frac{1}{2} = \frac{1 \times 3}{2 \times 3} = \frac{3}{6} \quad \text{and} \quad \frac{2}{3} = \frac{2 \times 2}{3 \times 2} = \frac{4}{6}$$

in order to express each fraction in terms of the denominator 6. The sum now becomes

$$\frac{3}{6} + \frac{4}{6} = \frac{7}{6}$$

since we can now simply add the numerators.

Now consider how we can perform the subtraction

$$\frac{a}{b} - \frac{c}{d}$$

In this case we need to realise that

$$\frac{a}{b} = \frac{a \times d}{b \times d} \quad \text{and} \quad \frac{c}{d} = \frac{c \times b}{d \times b}$$

Since $b \times d = d \times b$, these fractions now have the same denominator, which is obtained by multiplying the individual denominators together. The difference of the numerators can now be taken to give

$$\frac{a \times d}{b \times d} - \frac{c \times b}{d \times b} = \frac{(a \times d) - (c \times b)}{b \times d}$$

This can be written rather more neatly using implied multiplication as

$$\frac{ad - bc}{bd}$$

Multiplication

Multiplication of fractions is much more straightforward. If we wish to calculate

$$\frac{3}{4} \times \frac{2}{3}$$

we simply multiply the numerators and denominators separately to give

$$\frac{3 \times 2}{4 \times 3} = \frac{6}{12}$$

This can be simplified by dividing both numerator and denominator by 6 to give $1/2$.

In symbols we would have, for example

$$\frac{a}{b} \times \frac{c}{d} = \frac{a \times c}{b \times d} = \frac{ac}{bd}$$

Division

To divide one fraction by another, simply multiply by the reciprocal. So if we wish to calculate

$$\frac{4}{7} \div \frac{2}{3}$$

we would rewrite this as

$$\frac{4}{7} \times \frac{3}{2} = \frac{4 \times 3}{7 \times 2} = \frac{12}{14}$$

which simplifies, on dividing both numerator and denominator by 2, to 6/7.

An example using symbols (expressed slightly differently) is

$$\frac{a/b}{c/d} = \frac{a}{b} \times \frac{d}{c} = \frac{ad}{bc}$$

Empirical formulae

Titanium and oxygen combine in a certain ratio to give a compound which is used as a white pigment in paint. This can be obtained by combining 3.00 g of titanium with every 2.00 g of oxygen. The atomic masses of titanium and oxygen are 47.88 and 16.00 g mol^{-1} respectively.

We need to find the ratio of the number of atoms of each element in order to determine the empirical formula. This is equivalent to determining the ratios of the amount expressed in moles. Amount n can be found from the formula

$$n = \frac{m}{M}$$

where m is the mass of the element and M its molar mass. The ratio $n_{\text{Ti}}/n_{\text{O}}$ is thus given by the equation

$$\frac{n_{\text{Ti}}}{n_{\text{O}}} = \frac{m_{\text{Ti}}/M_{\text{Ti}}}{m_{\text{O}}/M_{\text{O}}}$$

which as we have seen above can be rearranged to give

$$\frac{n_{\text{Ti}}}{n_{\text{O}}} = \frac{m_{\text{Ti}}}{M_{\text{Ti}}} \times \frac{M_{\text{O}}}{m_{\text{O}}}$$

Substituting values from above gives

$$\frac{n_{\text{Ti}}}{n_{\text{O}}} = \frac{3.00\,\text{g}}{47.88\,\text{g mol}^{-1}} \times \frac{16.00\,\text{g mol}^{-1}}{2.00\,\text{g}} = 0.50$$

since the units within the fraction all cancel.

Since $n_{\text{Ti}}/n_{\text{O}} = 0.50$, we can take the reciprocal of each side to give $n_{\text{O}}/n_{\text{Ti}} = 1/0.50 = 2$, so the empirical formula of the compound is TiO_2.

The hydrogen atom spectrum

The wavenumbers $\tilde{\nu}$ of lines in the spectrum of the hydrogen atom are given by the equation

$$\tilde{\nu} = R\left(\frac{1}{m^2} - \frac{1}{n^2}\right)$$

where R is the Rydberg constant, m is an integer value of 1 or greater, and n is an integer value greater than m. The wavenumber is equal to $1/\lambda$, so the wavelength λ can be found by taking the reciprocal of $\tilde{\nu}$. Before we can do this it is necessary to rewrite the expression as a single fraction, rather than the difference between two fractions as at present. To do this we need a common denominator.

Taking the first term and multiplying by n^2 top and bottom gives

$$\frac{1}{m^2} = \frac{1 \times n^2}{m^2 \times n^2} = \frac{n^2}{m^2 n^2}$$

Multiplying the second term by m^2 top and bottom gives

$$\frac{1}{n^2} = \frac{1 \times m^2}{n^2 \times m^2} = \frac{m^2}{m^2 n^2}$$

The overall expression now becomes

$$\frac{1}{\lambda} = R\left(\frac{n^2}{m^2 n^2} - \frac{m^2}{m^2 n^2}\right)$$

which can easily be rewritten as

$$\frac{1}{\lambda} = R\left(\frac{n^2 - m^2}{m^2 n^2}\right)$$

since there is now a common denominator. As this is a single fraction, taking the reciprocal is straightforward and gives

$$\lambda = \frac{1}{R}\left(\frac{m^2 n^2}{n^2 - m^2}\right)$$

Questions

1. Evaluate the following and express them in their simplest form:
   (a) $\frac{1}{3} + \frac{1}{6}$
   (b) $\frac{3}{4} + \frac{2}{3}$
   (c) $\frac{2}{3} + \frac{1}{8}$
   (d) $\frac{2}{3} - \frac{1}{4}$
   (e) $\frac{4}{3} - \frac{3}{16}$

2. Evaluate the following and express the answer in its simplest form:
   (a) $\frac{1}{2} \times \frac{3}{4}$
   (b) $\frac{3}{8} \times \frac{3}{4}$
   (c) $\frac{1}{4} \times \frac{22}{7}$
   (d) $\frac{2}{3} \div \frac{3}{16}$
   (e) $\frac{1}{2} \div \frac{3}{4}$

3. Determine the empirical formula of a compound which contains 33.4 g of sulphur and 50.1 g of oxygen. The molar masses of sulphur and oxygen are 32.1 g mol^{-1} and 16.0 g mol^{-1} respectively.

4. In the hydrogen atom spectrum, calculate $\tilde{\upsilon}$ if $m = 1$ and $n = 3$ using the equation given in the text. Take the value of the Rydberg constant, R, from Appendix 4.

5. The ionisation energy of the hydrogen atom can be found by setting $m = 1$ and n to infinity in the expression for $\tilde{\upsilon}$. By considering the value of $1/n$ in this case, determine the value of $\tilde{\upsilon}$ which corresponds to this energy.

13. Inequalities

There are four types of inequalities:

1. 'Greater than', which is represented in symbols by '>'. Thus we might have a statement such as $x > 3$, which means that x can take any value as long as it is greater than 3. Notice that x can't be 3 exactly, but it could be very close to 3, for example 3.000 000 1.
2. 'Greater than or equal to', which is represented in symbols by '$\geq$'. This time the statement $x \geq 3$ does allow x to be exactly 3 or any value above.
3. 'Less than', which is represented in symbols by '<'. A statement such as $y < 5$ means that y can take any value as long as it is less than 5 exactly, such as 4.2 or even 4.999 999 999.
4. 'Less than or equal to', which is represented in symbols by '$\leq$'. In this case the statement $y \leq 5$ would allow y to take values 1.8, 3.2 or exactly 5, for example.

Solving inequalities

To solve an inequality such as $x + 2 < 9$, we manipulate the expression in exactly the same way as an equation, retaining the inequality sign and performing the same operations on either side of the inequality. In this case, subtracting 2 from either side gives

$$x + 2 - 2 < 9 - 2 \quad \text{or} \quad x < 7$$

Similarly, if $3x > 27$, dividing both sides by 3 gives

$$\frac{3x}{3} > \frac{27}{3} \quad \text{or} \quad x > 9$$

Multiplying by a negative number

If we have an expression such as $5 < 8$, which is clearly correct, it should be fairly obvious that multiplying this expression on both sides by -1 gives the expression $-5 < -8$, which is clearly untrue. It is thus important to remember that if an equality is multiplied throughout by a negative number we need to change $>$ to $<$, $<$ to $>$, $\leq$ to $\geq$ and $\geq$ to $\leq$.

Predicting a spontaneous reaction

The change ΔG of Gibbs free energy for a reaction can be found from the equation

$$\Delta G = \Delta H - T\Delta S$$

The value of ΔG allows us to predict whether a reaction will take place spontaneously or not, or whether it is at equilibrium. The conditions are

$$\begin{aligned} \Delta G &< 0 \quad \text{for a spontaneous reaction} \\ \Delta G &= 0 \quad \text{at equilibrium} \\ \Delta G &> 0 \quad \text{for a non-spontaneous reaction} \end{aligned}$$

Using the equations above we see that for a spontaneous reaction $\Delta G < 0$, so

$$\Delta H - T\Delta S < 0$$

Adding $T\Delta S$ to both sides of the equation gives

$$\Delta H < T\Delta S$$

For the reaction

$$4\,Ag_{(s)} + O_{2(g)} \rightleftharpoons 2\,Ag_2O_{(s)}$$

under standard conditions at 298 K, the value of ΔH is -62.10 kJ mol^{-1} and that of ΔS is -132.74 J K^{-1} mol^{-1}. Consequently $T\Delta S = 298\text{ K} \times (-132.74\text{ J K}^{-1}\text{ mol}^{-1})$, which is $-39\,556$ J mol^{-1}. Dividing this by 1000 gives $T\Delta S = -39.6$ kJ mol^{-1}. Since

$$-62.10\text{ kJ mol}^{-1} < -39.6\text{ kJ mol}^{-1}$$

is a true statement, the reaction is spontaneous under these conditions.

The variation theorem

One of the postulates of quantum mechanics tells us that the physical state of a particle can be fully described by an appropriate mathematical function which is known as its wavefunction. The variation theorem states that any wavefunction which obeys the same boundary conditions as the correct ground state wavefunction will give an expectation value $\bar{E}$ for the energy that is greater than or equal to the true ground state energy E_o, i.e. $\bar{E} \geq E_0$.

The particle in a box model can be used to illustrate many of the techniques of quantum mechanics in chemistry. It is also of some use in predicting the absorption spectra of delocalised systems such as hexatriene. For the particle in a box model the true ground state energy is given by

$$E_0 = \frac{h^2}{8ma^2}$$

while a particular trial wavefunction gives the expectation value $\bar{E}$ of the energy as

$$\bar{E} = \frac{6h^2}{4\pi^2 ma^2}$$

In these equations, h is Planck's constant, m the mass of the particle and a the length of the box in which the particle is confined. This latter expression can be written as

$$\bar{E} = \frac{12h^2}{8\pi^2 ma^2} = \left(\frac{12}{\pi^2}\right)\frac{h^2}{8ma^2} = 1.217\frac{h^2}{8ma^2}$$

Since

$$1.217\frac{h^2}{8ma^2} > \frac{h^2}{8ma^2} \quad \text{or} \quad 1.217 > 1$$

it follows that $\bar{E} > E_0$ as expected.

Questions

1. Solve the following inequalities for the given variable:
   (a) $x + 4 > 13$
   (b) $y + 7 > 25$
   (c) $x - 3 < 10$
   (d) $x - 10 \leq 16$
   (e) $5 + y \geq 17$

2. Solve the following inequalities for the given variable:
   (a) $9 - x > 2$
   (b) $4 - x < 3$
   (c) $2 - y < -6$
   (d) $14 - x \geq 7$
   (e) $6 - y \leq 1$

3. Solve the following inequalities for the given variable:
   (a) $3x > 18$
   (b) $4x + 2 < 18$
   (c) $9 - 3x \geq 72$
   (d) $3y - 7 \leq 28$
   (e) $5 - 4x \leq 29$

4. For the one-dimensional particle in a box, the wavefunction $\Psi_1 = a(a - x)$ gives an energy $E_1 = 1.013h^2/8ma^2$. A second wavefunction $\Psi_2 = B\sin\pi x/a$ gives energy $E_2 = h^2/8ma^2$. Which of these is the true wavefunction?

5. For the reaction

$$N_{2(g)} + 3\,H_{2(g)} \rightarrow 2\,NH_{3(g)}$$

under standard conditions at 298 K, $\Delta H = -91.8$ kJ mol^{-1} and $\Delta S = -197$ J K^{-1} mol^{-1}. By calculating ΔG, determine whether or not the reaction is spontaneous.

14. Rearranging Equations

It is often necessary to rearrange an equation, whether to solve it or for some other purpose. There is only one rule to remember when doing so: perform exactly the same operation on each side. So, for example, if we add 4 to the left-hand side we must add 4 to the right-hand side.

Consider the very simple equation

$$x + 2 = 5$$

It is fairly obvious that $x = 3$, but subtracting 2 from either side gives

$$x + 2 - 2 = 5 - 2$$

which reduces to $x = 3$.

A more complicated example is

$$\frac{x-2}{3} = 4$$

It is generally necessary to remove any fraction first. Do this by multiplying both sides by 3

$$\left(\frac{x-2}{3}\right) \times 3 = 4 \times 3 \quad \text{or} \quad x - 2 = 12$$

Then add 2 to each side to give $x = 14$.

As a final example, consider the expression

$$\surd(x - y) = \frac{a+b}{c}$$

which we wish to transform to make x the subject, i.e. an expression of the form 'x equals . . . '. First remove the square root by squaring each side of the equation

$$(\surd(x - y))^2 = \left(\frac{a+b}{c}\right)^2$$

or more simply

$$x - y = \left(\frac{a+b}{c}\right)^2$$

To leave x on its own, simply add y to each side

$$x - y + y = \left(\frac{a+b}{c}\right)^2 + y$$

which gives

$$x = \left(\frac{a+b}{c}\right)^2 + y$$

Dimerisation of benzoic acid

In benzene the molecules of benzoic acid C_6H_5COOH form a dimer. The equilibrium can be represented by the equation

$$(C_6H_5COOH)_2 \rightleftharpoons 2\,C_6H_5COOH$$

with an associated equilibrium constant K defined by

$$K = \frac{[C_6H_5COOH]^2}{[(C_6H_5COOH)_2]}$$

which relates the concentrations of the reactants and the products once equilibrium has been reached. To obtain the concentration of C_6H_5COOH, begin by multiplying both sides of the equation by

$[(C_6H_5COOH)_2]$ to give

$$K\,[(C_6H_5COOH)_2] = [C_6H_5COOH]^2$$

Then take the square root of each side

$$\sqrt{(K\,[(C_6H_5COOH)_2])} = \sqrt{([C_6H_5COOH]^2)}$$

which becomes

$$[C_6H_5COOH] = \sqrt{(K\,[C_6H_5COOH]^2)}$$

Raoult's law

Above a mixture of two liquids the vapour pressure p is given by the expression

$$p = x_1 p_1^* + x_2 p_2^*$$

where x_1 and x_2 are the mole fractions of the two components in the mixture, and p_1^* and p_2^* are the respective vapour pressures of the pure components.

To obtain the mole fraction x_2, first subtract the term $x_1 p_1^*$ from each side to give

$$p - x_1 p_1^* = x_1 p_1^* + x_2 p_2^* - x_1 p_1^* \quad \text{or} \quad p - x_1 p_1^* = x_2 p_2^*$$

Then divide each side by p_2^* to give

$$\frac{p - x_1 p_1^*}{p_2^*} = \frac{x_2 p_2^*}{p_2^*} \quad \text{or} \quad x_2 = \frac{p - x_1 p_1^*}{p_2^*}$$

Questions

1. Rearrange the following equations to make y the subject:
   (a) $x + y = 4$
   (b) $3x + 2y = 17$
   (c) $x^2 - y^2 = 5$
   (d) $4x^2 y = 20$
   (e) $3x^2 y^2 + 2 = 19$

2. Rearrange the ideal gas equation, $pV = nRT$, to make p, V, n and T in turn the subject. In this equation p represents pressure, V volume, n the amount of gas, R the gas constant and T the absolute temperature.

3. The phase rule

$$P + F = C - 2$$

   relates the number of phases in a system with C components to the degrees of freedom F. Rearrange this equation to give an expression for F.

4. The change in Gibbs free energy, ΔG, is related to the changes in enthalpy, ΔH, and entropy, ΔS, by the equation

$$\Delta G = \Delta H - T\Delta S$$

   where T is the absolute temperature. Obtain an expression for ΔS in general and when $\Delta G = 0$.

5. In the steady-state approximation of the analysis of the kinetics of the reaction for the decomposition of ethane, the equation

$$k_1[C_2H_6] - k_2[CH_3][C_2H_6] = 0$$

   is obtained. Rearrange this equation to give an expression for $[CH_3]$.

6. As given in Chapter 12, the wavenumbers, $\tilde{\nu}$, of lines in the hydrogen atom are given by the equation

$$\tilde{\nu} = R\left(\frac{1}{n_1^2} - \frac{1}{n_2^2}\right)$$

   where R is the Rydberg constant and n_1 and n_2 are integers. Obtain an expression for n_2 from this equation.

15. Ratios and Proportionality

The ratio of two numbers is simply the fraction obtained when one is divided by the other. The ratio of 3 and 4 is consequently 3/4 or 0.75; alternatively we may choose to use the ratio of 4/3 if this is more appropriate.

The concept of proportionality will be familiar to most people. It is logical to expect that if we double the volume of a substance then its mass will also double. In this case we describe the relationship as direct proportionality.

Mathematically, we define two quantities as being in direct proportion if their ratio is constant. This would be the case for the two quantities x and y which take the following values:

x	2	4	6	8	10	12
y	3	6	9	12	15	18

Here we see that the ratio y/x has a constant value of 1.5. This simple equation can be arranged to give $y = 1.5x$, where we call 1.5 the constant of proportionality. Another way of defining direct proportionality is to say that the two quantities must obey a relationship of the form $y = kx$. This can also be written using the proportionality symbol as $y \propto x$.

Inverse proportionality may be less obvious, although it still frequently occurs in everyday life. A common example is the time taken to cover a fixed distance, which is inversely proportional to the speed of travel.

In mathematical terms, two quantities x and y would be inversely proportional if they had values such as:

x	5	10	15	20	25	30
y	30	15	10	7.5	6	5

In this case there is a constant product, so that $xy = 150$ in each case. More generally, x and y are said to be inversely proportional if $xy = k$, where k is again the proportionality constant. Using the proportionality symbol this can be written as $y \propto 1/x$.

Beer–Lambert law

The Beer–Lambert law can be expressed using the equation

$$A = \varepsilon cl$$

and gives the absorbance A of light passing through solution of concentration c for a distance l. The constant ε is known as the coefficient of extinction and is constant for a given material; for a given sample, l will also be constant.

If we define k as the product εl, the equation becomes

$$A = kc$$

and we see that the absorbance is directly proportional to concentration.

This relationship can be used in kinetic experiments where absorbance (measured with a spectrophotometer) can be monitored rather than concentration directly.

The ideal gas equation

The well-known ideal gas equation

$$pV = nRT$$

gives the pressure p of an amount n of ideal gas in terms of its volume V at temperature T. The quantity R is known as the gas constant.

For a fixed amount n of gas at constant temperature T, the three-constant term can be grouped and defined as $k = nRT$. The equation then becomes

$$pV = k$$

and it can be seen that the pressure and volume are inversely proportional.

Questions

1. Determine the nature of the relationship between x and y, including the value of any proportionality constant, from the following data:

x	0.25	0.50	0.75	1.00	1.25	1.50	1.75	2.00
y	1.00	2.00	3.00	4.00	5.00	6.00	7.00	8.00

2. Determine the nature of the relationship between x and y, including the value of any proportionality constant, from the following data:

x	0.50	1.00	1.50	2.00	2.50	3.00	3.50	4.00	4.50	5.00
y	10.0	5.00	3.33	2.50	2.00	1.67	1.43	1.25	1.11	1.00

3. Identify the nature of any proportionality relationships between x, y and z, giving the value of any proportionality constant, from the following data:

x	1.00	2.00	3.00	4.00	5.00	6.00	7.00	8.00	9.00	10.00
y	3.00	6.00	9.00	12.00	15.00	18.00	21.00	24.00	27.00	30.00
z	1.00	0.50	0.33	0.25	0.20	0.167	0.143	0.125	0.111	0.100

4. The force F acting on a bond within a molecule is directly proportional to its displacement x from its usual length. A displacement of 5×10^{-12} m in Cl_2 results in a force of 1.6×10^{-9} N. What is the constant of proportionality?

5. Wavelength is inversely proportional to frequency. A wavelength of 510 nm corresponds to a frequency of 5.88×10^{14} Hz. Determine the constant of proportionality.

6. Cyclopropane isomerises to propene at 1000°C. The rate is 0.92 s^{-1} when the concentration of cyclopropane is 0.1 mol dm^{-3}, and 0.46 s^{-1} when the concentration of cyclopropane is 0.05 mol dm^{-3}. Deduce the relationship between rate and cyclopropane concentration, giving the value of any constants.

16. Factorials

The factorial of a number is simply the product of itself and each integer reducing by 1. For example, the factorial of 5 is

$$5 \times 4 \times 3 \times 2 \times 1$$

In symbols the factorial of a number n is written as $n!$ Consequently the factorial of 5 as defined above can be denoted as 5!

Special cases

It is probably obvious that $1! = 1$, but less so that $0! = 1$.

Simplifying factorial expressions

A little thought shows that the definition of 5! given above can also be written as

$$5 \times 4! \quad \text{or} \quad 5 \times 4 \times 3!$$

since $4! = 4 \times 3 \times 2 \times 1$ and $3! = 3 \times 2 \times 1$.
This can be of use in simplifying expressions involving quotients of factorials such as

$$\frac{8!}{6!} = \frac{8 \times 7 \times 6!}{6!} = 8 \times 7 = 56$$

Nuclear magnetic resonance

This is a spectroscopic technique which allows the number of protons within each group in a molecule to be identified. For example, the NMR spectrum of ethanol CH_3CH_2OH would consist of three groups of protons, from the CH_3, CH_2 and OH groups. The signal due to each group of protons is split, according to the number of protons in an adjacent group. The general rule is that n protons in an adjacent group result in a splitting of the signal into $n + 1$ separate peaks. In this case the signal due to CH_3 will be split into 3 by the 2 protons on the adjacent group.

The relative intensities of such split peaks can be determined using a formula that involves factorials. In the case above, the signal due to CH_2 will be split into 4 by the 3 protons in the CH_3 group. The intensities of these four peaks are represented in symbols as $^3C_0$, $^3C_1$, $^3C_2$ and $^3C_3$. To calculate the relative intensity of each, use the general formula

$$^nC_r = \frac{n!}{(n-r)!\,r!}$$

and substitute for n and r for each term. In the first one, $n = 3$ and $r = 0$ so

$$^3C_0 = \frac{3!}{(3-0)!\,0!} = \frac{3!}{3! \times 0!} = \frac{1}{0!} = \frac{1}{1} = 1$$

The second term has $n = 3$ and $r = 1$, giving

$$^3C_1 = \frac{3!}{(3-1)!\,1!} = \frac{3 \times 2!}{2! \times 1} = \frac{3}{1} = 3$$

using the fact that 3! can be written as $3 \times 2!$, as described previously.

The third term has $n = 3$ and $r = 2$, so

$$^3C_2 = \frac{3!}{(3-2)!\,2!} = \frac{3 \times 2!}{1! \times 2!} = \frac{3}{1} = 3$$

Finally term four has $n = 3$ and $r = 3$, giving

$$^{3}C_3 = \frac{3!}{(3-3)!\,3!} = \frac{3!}{0! \times 3!} = \frac{1}{0!} = \frac{1}{1} = 1$$

The relative intensities of the four peaks are therefore 1, 3, 3, 1.

Statistical mechanics

In summary, this is an area of physical chemistry where statistical methods are applied to the quantum model of atoms and molecules and used to calculate macroscopic thermodynamic quantities.

One of the problems which arises in statistical mechanics is how to distribute N molecules across a number of energy levels. Typically we would say that n_0 molecules occupy level 0 with energy ε_0, n_1 molecules occupy level 1 with energy ε_1, n_2 molecules occupy level 2 with energy ε_2, and so on, so that in general n_i molecules occupy energy level i with energy ε_i. The number of ways in which a given distribution can occur is known as the number Ω of complexions, and is given by the formula

$$\Omega = \frac{N!}{\Pi_i n_i!}$$

This introduces the symbol Π, which simply means take the product of the following quantities. So in this case $\Pi_i n_i!$ means $n_0! \times n_1! \times n_2! \times \ldots \times n_i!$.

As a specific example, consider the case where for the first three levels 4 molecules have energy ε_o, 2 molecules have energy ε_1 and 3 molecules have energy ε_2. Thus $n_o = 4$, $n_1 = 2$ and $n_2 = 3$, with the total N being $4 + 2 + 3 = 9$. The total number of ways in which this arrangement can be achieved is

$$\Omega = \frac{9!}{4!2!3!}$$

This can be evaluated directly using a calculator, or by simplifying in successive steps to give

$$\Omega = \frac{9!}{4!2!3!} = \frac{9 \times 8 \times 7 \times 6 \times 5 \times 4!}{4!2!3!} = \frac{9 \times 8 \times 7 \times 6 \times 5}{2 \times 1 \times 3 \times 2 \times 1} = \frac{9}{3} \times \frac{8}{2} \times 7 \times \frac{6}{2} \times 5$$

$$= 3 \times 4 \times 7 \times 3 \times 5 = 1260$$

Questions

1. Evaluate the following without using a calculator:
   (a) 3!
   (b) 4!
   (c) 5!
   (d) 6!
   (e) 7!

2. Evaluate the following by using a calculator:
   (a) 10!
   (b) 12!
   (c) 15!
   (d) 25!
   (e) 36!

3. Simplify the following expressions:
   (a) $\frac{8!}{6!}$
   (b) $\frac{5!}{4!}$
   (c) $\frac{20!}{17!}$
   (d) $\frac{10!}{8!} \times \frac{5!}{7!}$
   (e) $\frac{36!}{30!}$

4. Evaluate the following terms:
   (a) ${}^{4}C_2$
   (b) ${}^{5}C_3$
   (c) ${}^{5}C_4$
   (d) ${}^{8}C_6$
   (e) ${}^{8}C_7$

5. Give the splitting patterns and the relative intensities of the peaks in the compound $CH_3CH_2CH_3$.

6. A system containing 10 molecules has 4 molecules with energy ε_o, 3 molecules with energy ε_1, 2 molecules with energy ε_2 and 1 molecule with energy ε_3. In how many ways can this arrangement be achieved?

Functions

17. Functions of a Single Variable

When the value of a quantity depends on that of a second quantity, it is said to be a function of the second quantity or variable. For example, the mass of a standard piece of A4 paper will be a function of its thickness.

In mathematical notation, we would describe a function f of a single variable x as $f(x)$. The function can then be defined by setting $f(x)$ equal to some expression in x. An example would be

$$f(x) = x + 3$$

In this case the function is obtained by taking the variable x and adding 3 to it. Clearly the function in this case will have a different value for every value of x. To obtain the value of a function for a given value of x, we simply replace x by that variable in the above expression. Thus if $x = 2$, we substitute directly to give

$$f(2) = 2 + 3 = 5$$

so that 2 appears in place of x everywhere in the defining statement.

Functions can be more complicated than this. A second example is

$$f(x) = x^2 + \frac{1}{x}$$

in which setting $x = 2$ gives

$$f(2) = 2^2 + \frac{1}{2} = 4 + \frac{1}{2} = \frac{8}{2} + \frac{1}{2} = \frac{9}{2}$$

Also note that $f(0)$ isn't defined since this would involve the calculation of 1/0.

Heat capacity

The heat capacity, C_p, of nitrogen can be expressed in terms of absolute temperature T in the form

$$C_p(T) = a + bT + cT^{-2}$$

where a, b and c are constants which have the values

$$a = 28.58\,\mathrm{J\,K^{-1}\,mol^{-1}}$$
$$b = 3.76 \times 10^{-3}\,\mathrm{J\,K^{-2}\,mol^{-1}}$$
$$c = -5.0 \times 10^{4}\,\mathrm{J\,K\,\,mol^{-1}}$$

Thus the heat capacity of nitrogen at any temperature can be calculated. At room temperature, $T = 298$ K, and so

$$\begin{aligned} C_p(298\,\mathrm{K}) &= 28.58\,\mathrm{J\,K^{-1}\,mol^{-1}} + (3.76 \times 10^{-3}\,\mathrm{J\,K^{-2}\,mol^{-1}} \times 298\,\mathrm{K}) \\ &\quad + (-5.0 \times 10^{4}\,\mathrm{J\,K\,mol^{-1}} \times (298\,\mathrm{K})^{-2}) \\ &= 28.58\,\mathrm{J\,K^{-1}\,mol^{-1}} + 1.12\,\mathrm{J\,K^{-1}\,mol^{-1}} - 0.56\,\mathrm{J\,K^{-1}\,mol^{-1}} \\ &= 29.14\,\mathrm{J\,K^{-1}\,mol^{-1}} \end{aligned}$$

Notice how for each term the units combine so that we are adding three terms in $\mathrm{J\,K^{-1}\,mol^{-1}}$.

Particle in a box

As we saw in Chapter 13, a one-dimensional box is a model often used in introductory quantum mechanics. The energy E of a particle of mass m moving in a box of width a is defined by the function $E(n)$ where

$$E(n) = \frac{n^2h^2}{8\,ma^2}$$

and n is the quantum number which defines the energy levels. In this expression, h represents Planck's constant. In this function n is only allowed to take integer values from 1 upwards. This expression can be used to determine the difference between two energy levels, such as those define by $n = 3$ and $n = 4$. This difference will be given by

$$E(4) - E(3)$$

Substituting for n in the defining equation gives this difference as

$$\frac{4^2h^2}{8ma^2} - \frac{3^2h^2}{8ma^2} = \frac{16h^2}{8ma^2} - \frac{9h^2}{8ma^2} = \frac{7h^2}{8ma^2}$$

This can be converted to a numerical value for an electron of mass 9.1×10^{-31} kg confined in a one-dimensional box of length 3.5×10^{-10} m. Planck's constant h can be taken as 6.63×10^{-34} J s (Appendix 4). The energy difference will thus be

$$\frac{7 \times (6.63 \times 10^{-34}\,\text{J s})^2}{8 \times 9.1 \times 10^{-34}\,\text{kg} \times (3.5 \times 10^{-10}\,\text{m})^2} = \frac{7 \times 43.96 \times 10^{-68}\,\text{J}^2\,\text{s}^2}{8 \times 9.1 \times 10^{-34}\,\text{kg} \times 12.25 \times 10^{-20}\,\text{m}^2}$$

This can now be simplified. The overall power of 10 will be $-68 - (-34 - 20) = -68 - (-54) = -68 + 54 = -14$. The remaining numbers can be grouped and evaluated to give

$$\frac{7 \times 43.96}{8 \times 9.1 \times 12.25} = 0.345$$

The units can be simplified by realising that $1\,\text{J} = 1\,\text{kg m}^2\,\text{s}^{-2}$ (Appendix 2). The units then combine to give

$$\frac{(\text{kg m}^2\,\text{s}^{-2})^2\,\text{s}^2}{\text{kg m}^2} = \frac{\text{kg}^2\,\text{m}^4\,\text{s}^{-4}\,\text{s}^2}{\text{kg m}^2} = \text{kg m}^2\,\text{s}^{-2} = \text{J}$$

The overall answer is thus 0.345×10^{-14} J, or expressed slightly better as 3.45×10^{-15} J.

Questions

1. A function $f(x)$ is defined as $f(x) = 3x - 4$. Evaluate:
   (a) $f(-2)$
   (b) $f(0)$
   (c) $f(3)$

2. A function $f(x)$ is defined by $f(x) = 4x^2 - 2x - 6$. Evaluate:
   (a) $f(-3)$
   (b) $f(0)$
   (c) $f(2)$
   (d) $f(1/2)$

3. A function $g(y)$ is defined by

$$g(y) = \frac{1}{y} + \frac{2}{y^2} + \frac{3}{y^3}$$

   Evaluate:
   (a) $g(-2)$
   (b) $g(1/4)$
   (c) $g(4)$

   When is $g(y)$ undefined?

4. The relative density ρ of a solution of alcohol in water can be defined in terms of the mole fraction x of alcohol, i.e. in terms of the function $\rho(x)$. This function is defined by the equation

$$\rho(x) = 0.987 - 0.269x + 0.304x^2 - 0.598x^3$$

   Evaluate the relative density when $x = 0.15$.

5. The Lennard–Jones potential $V(r)$ gives the potential energy V between two molecules in terms of their separation r. It can be defined by the equation

$$V(r) = 4\varepsilon\left[\left(\frac{\sigma}{r}\right)^{12} - \left(\frac{\sigma}{r}\right)^{6}\right]$$

where ε and σ are constants that depend on the nature of the interacting molecules. For oxygen, $\varepsilon = 1.63 \times 10^{-21}$ J and $\sigma = 358$ pm. Determine $V(350 \text{ pm})$ for oxygen.

6. Raoult's law gives the value of the vapour pressure p above a mixture of two liquids which have vapour pressures $p_1{}^*$ and $p_2{}^*$ respectively. If the mole fraction of component 1 is x_1 then

$$p(x_1) = x_1\, p_1{}^* + (1 - x_1)\, p_2{}^*$$

Calculate $p(0.300)$ at 100°C for a mixture of benzene (component 1) and toluene (component 2) with $p_1{}^* = 1.800 \times 10^5$ Pa and $p_2{}^* = 0.742 \times 10^5$ Pa.

18. Functions of Many Variables

It is probably obvious that in most cases the dependence between quantities is much more complicated than that of a function of a single variable. Although this chapter is concerned with functions of many variables, we will illustrate these by considering a function f of two variables x and y. The extension to further variables is then straightforward.

In mathematical notation the function referred to above would be denoted as $f(x, y)$. In the example where this is defined as

$$f(x, y) = 3x^2y - 4xy + 2y - 8$$

we would have

$$\begin{aligned} f(2, 1) &= (3 \times 2^2 \times 1) - (4 \times 2 \times 1) + (2 \times 1) - 8 \\ &= (3 \times 4 \times 1) - (4 \times 2 \times 1) + (2 \times 1) - 8 \\ &= 12 - 8 + 2 - 8 \\ &= -2 \end{aligned}$$

where x has been replaced by 2 and y has been replaced by 1.

Similarly,

$$\begin{aligned} f(0, -1) &= (3 \times 0^2 \times (-1)) - (4 \times 0 \times (-1)) + (2 \times (-1)) - 8 \\ &= 0 + 0 - 2 - 8 \\ &= -10 \end{aligned}$$

Note that if one variable is fixed, say $x = 3$, we have

$$\begin{aligned} f(3, y) &= (3 \times 3^2 \times y) - (4 \times 3 \times y) + 2y - 8 \\ &= (3 \times 9 \times y) - (4 \times 3 \times y) + 2y - 8 \\ &= 27y - 12y + 2y - 8 \\ &= 17y - 8 \end{aligned}$$

which is a function of the single variable y.

Vibration–rotation spectroscopy

The Born–Oppenheimer approximation states the vibrational and rotational energies of a molecule can be separated and the individual terms added. The overall energy E can thus be considered as a function of the vibrational quantum number v (taking values 0, 1, 2, ...) and the rotational quantum number J (which independently takes values 0, 1, 2, ...). For a diatomic molecule it can be approximated by the function

$$E(v, J) = \left(v + \frac{1}{2}\right) h\nu_0 + BJ(J + 1)h$$

where B is the rotational constant of the molecule and h is Planck's constant.

Selection rules tell us which transitions between rotational and vibrational levels are allowed. The values of v may increase or decrease by 1. Similarly the values of J may independently increase or decrease by 1. For a transition between vibrational levels v'' and v', where $v' - v'' = 1$, we have

$$E(v'', J'') = \left(v'' + \frac{1}{2}\right) h\nu_0 + BJ''(J'' + 1)h$$

and

$$E(v', J') = \left(v' + \frac{1}{2}\right) h\nu_0 + BJ'(J' + 1)h$$

so that

$$E(v', J') - E(v'', J'') = (v' - v'')h\nu_0 + B[J'(J' + 1) - J''(J'' + 1)]h$$

denoting the rotational level in the upper vibrational state as J' and that in the lower vibrational state as J''. Since $v' - v'' = 1$, this simplifies to

$$E(v', J') - E(v'', J'') = h\nu_0 + B[J'(J' + 1) - J''(J'' + 1)]h$$

If $J' = J'' + 1$, the square bracket becomes

$$\left[(J'' + 1)(J'' + 2) - J''(J'' + 1)\right] = J''^2 + J'' + 2J'' + 2 - J''^2 - J'' = 2J'' + 2$$

This can also be written as $2(J'' + 1)$, and so

$$E(v', J') - E(v'', J'') = h\nu_0 + 2Bh(J'' + 1) = h\nu_0 + 2BhJ'$$

Since J' can take only integral values these energy levels are equally spaced. The resulting branch of the spectrum is known as the R branch.

Three-dimensional harmonic oscillator

The potential energy V of a three-dimensional harmonic oscillator depends on the position of a particle in space, as described by the three coordinates x, y and z. It can thus be considered to be a function of these variables written as $V(x, y, z)$ and defined by the equation

$$V(x, y, z) = \frac{1}{2}k_x x^2 + \frac{1}{2}k_y y^2 + \frac{1}{2}k_z z^2$$

where k_x, k_y and k_z are force constants in each of the three directions.

Consider a three-dimensional harmonic oscillator in which $k_x = 100\ \text{N m}^{-1}$, $k_y = 150\ \text{N m}^{-1}$, and $k_z = 250\ \text{N m}^{-1}$. The potential energy when $x = 0.5$ nm, $y = 0.7$ nm and $z = 0.9$ nm could be calculated by substituting in the expression above to give

$$\begin{aligned} V(0.5\,\text{nm}, 0.7\,\text{nm}, 0.9\,\text{nm}) &= \left(\frac{1}{2} \times 100\,\text{N mol}^{-1} \times (0.5\,\text{nm})^2\right) \\ &+ \left(\frac{1}{2} \times 150\,\text{N mol}^{-1} \times (0.7\,\text{nm})^2\right) \\ &+ \left(\frac{1}{2} \times 250\,\text{N mol}^{-1} \times (0.9\,\text{nm})^2\right) \end{aligned}$$

Remembering that 1 nm $= 10^{-9}$ m (Appendix 1) and consequently that 1 nm$^2 = 10^{-18}$ m^2 allows us to evaluate this as

$$\begin{aligned} V(0.5\,\text{nm}, 0.7\,\text{nm}, 0.9\,\text{nm}) &= (0.5 \times 100\,\text{N mol}^{-1} \times 0.25 \times 10^{-18}\,\text{m}^2) + (0.5 \times 150\,\text{N mol}^{-1} \\ &\quad \times 0.49 \times 10^{-18}\,\text{m}^2) + (0.5 \times 250\,\text{N mol}^{-1} \times 0.81 \times 10^{-18}\,\text{m}^2) \\ &= (12.5 \times 10^{-18}\,\text{N m}) + (36.75 \times 10^{-18}\,\text{N m}) + (101.25 \times 10^{-18}\,\text{N m}) \\ &= 150.5 \times 10^{-18}\,\text{N m} \end{aligned}$$

Since 1 N m = 1 J and 10^{-18} J is a commonly used unit for small energies known as an attojoule, aJ (see Appendix 1), the final answer can be written as

$$V(0.5\ \text{nm}, 0.7\ \text{nm}, 0.9\ \text{nm}) = 150.5\ \text{aJ}$$

Questions

1. A function $f(x, y)$ is defined by $f(x, y) = 1 + 2x - 3y$. Evaluate:
   (a) $f(1, 1)$
   (b) $f(0, 0)$
   (c) $f(-2, 0)$
   (d) $f(-3, -2)$
   (e) $f(0, 3)$

2. A function $g(x, y, z)$ is defined by $g(x, y, z) = 3x^2 - 4y + z$. Evaluate:
   (a) $g(1, 0, -1)$
   (b) $g(2, 2, 0)$
   (c) $g(-3, -2, 1)$
   (d) $g(-2, 0, 3)$
   (e) $g(-3, 4, -2)$

3. A function $f(x, y)$ is defined by $f(x, y) = 2x^2y - 3xy^2$. Evaluate:
   (a) $f(2, 1)$
   (b) $f(0, 3)$
   (c) $f(-2, 1)$
   (d) $f(1, -2)$
   (e) $f(-1, -2)$

4. The ideal gas equation can be rearranged and expressed as a function in T and V as

$$p(T,\ V) = \frac{nRT}{V}$$

where p is pressure, V volume, T absolute temperature, and n the amount of gas. R is the gas constant, 8.314 J K^{-1} mol^{-1}. Evaluate p (298 K, 1.5 m^3) for 2.5 mole of an ideal gas.

5. The energy E of a three-dimensional particle in a box can be expressed in terms of the three quantum numbers n_x, n_y and n_z as

$$E(n_x, n_y, n_z) = \frac{h^2}{8m}\left(\frac{n_x^2}{a^2} + \frac{n_y^2}{b^2} + \frac{n_z^2}{c^2}\right)$$

where a, b and c are the dimensions of the box in the x, y and z directions respectively, h is Planck's constant, and m is the mass of the particle. For an electron of mass 9.11×10^{-31} kg, evaluate E (1,2,1) for a box with $a = b = c = 200$ pm.

19. Natural Logarithms

If three numbers a, b and c are related by the equation

$$a = b^c$$

then c is said to be the logarithm of a to the base b. This would be expressed in terms of an equation as

$$c = \log_b a$$

Three numbers which obey such a relationship would be

$$8 = 2^3$$

so that

$$3 = \log_2 8$$

Logarithms can be taken to any base, which does not have to be an integer. In practice, there are only two bases which need to be considered for chemistry.

The first of these is the exponential number e, which has the value 2.718 to 3 decimal places. Logarithms to the base e are known as natural logarithms and denoted as ln; the natural logarithm of x is thus written as $\ln x$.

It is useful to construct a table of various logarithmic values

x	0.10	0.25	0.50	1.0	1.5	2.0	2.5	3.0	3.5	4.0
$\ln x$	−2.30	−1.39	−0.693	0.000	0.405	0.693	0.916	1.099	1.25	1.39

from which the graph shown in Fig. 19.1 can be constructed.

From the graph it can be seen that any fractional value of x leads to a negative value of $\ln x$. Also, negative values of x have deliberately been excluded and plugging these into a calculator immediately shows that $\ln x$ is not defined in such cases.

Rules of logarithms

These rules apply to logarithms taken to any base.

- If $a = b \times c$, then $\ln a = \ln b + \ln c$, i.e. if two numbers are multiplied, the logarithm of the result is obtained by adding the individual logarithms.

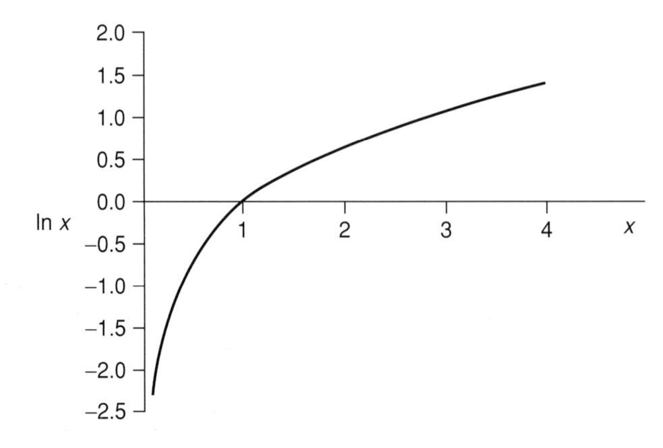

Figure 19.1 Graph of ln x against x

For example, since $12 = 3 \times 4$ then $\ln 12 = \ln 3 + \ln 4$. Since $\ln 3 = 1.099$ and $\ln 4 = 1.386$, we have $\ln 12 = 1.099 + 1.386 = 2.485$, which is correct.

- If $a = b/c$, then $\ln a = \ln b - \ln c$, i.e. if one number is divided by another, the logarithm of the result is obtained by taking the difference of the logarithms.

For example, since $5 = 10/2$ then $\ln 5 = \ln 10 - \ln 2$. Since $\ln 10 = 2.303$ and $\ln 2 = 0.693$, we have $\ln 5 = 2.303 - 0.693 = 1.610$, which is correct.

- If $a = b^c$, then $\ln a = c \ln b$, i.e. if one number is raised to a power, the logarithm of the result is obtained by multiplying the logarithm of that number by the power.

For example, since $9 = 3^2$, then $\ln 9 = 2 \ln 3$. Since $\ln 3 = 1.099$, we have $\ln 9 = 2 \times 1.099 = 2.198$, which is correct.

Nernst equation

This allows us to predict the electromotive force (EMF), E, of an electrochemical cell under a specified set of conditions. The EMF under standard conditions is easily predicted from tables of data and gives us the standard EMF usually denoted by $E^{\ominus}$. The relationship between these quantities is given by the equation

$$E = E^{\ominus} - \frac{RT}{nF} \ln Q$$

where R is the gas constant, T the absolute temperature in K, and F a constant known as the Faraday constant (Appendix 4).

The quantities n and Q are best introduced via a specific example. For the electrochemical cell overall reaction

$$Zn_{(s)} + Cu^{2+}{}_{(aq)} \rightleftharpoons Cu_{(s)} + Zn^{2+}{}_{(aq)}$$

we have

$$Q = \frac{[Zn^{2+}]}{[Cu^{2+}]}$$

which is the ratio of the concentrations of the products multiplied together and the concentrations of the reactants multiplied together, ignoring any solid species. In this case $n = 2$, which is the numerical amount of electrons (i.e. the 'number of moles', with no units) transferred in the reaction.

The standard electromotive force $E^{\ominus}$ for the above reaction is 1.10 V. For a cell in which $[Zn^{2+}] = 1.5 \times 10^{-5}$ mol dm^{-3} and $[Cu^{2+}] = 0.100$ mol dm^{-3}, substituting in the Nernst equation gives

$$\begin{aligned} E &= 1.10\,\text{V} - \frac{8.314\,\text{JK}^{-1}\,\text{mol}^{-1} \times 298\,\text{K}}{2 \times 96485\,\text{C}\,\text{mol}^{-1}} \ln\left(\frac{1.5 \times 10^{-5}\,\text{mol dm}^{-3}}{0.100\,\text{mol dm}^{-3}}\right) \\ &= 1.10\,\text{V} - 0.013\,\text{J}\,\text{C}^{-1} \times \ln\left(1.5 \times 10^{-4}\right) \end{aligned}$$

At this point it must be realised that 1 J C^{-1} = 1 V (Appendix 2), so that the units in this equation are consistent and

$$\begin{aligned} E &= 1.10\,\text{V} - (0.013\,\text{V} \times -8.80) \\ &= 1.10\,\text{V} + 0.114 \\ &= 1.21\,\text{V} \end{aligned}$$

giving the answer to an appropriate number of decimal places.

Entropy of expansion of an ideal gas

When an ideal gas expands at constant temperature its entropy S increases. This is consistent with the increase in disorder which occurs as the gas molecules are able to move over larger distances. If the gas expands from volume V_1 to volume V_2, the entropy change ΔS is given by

$$\Delta S = nR \ln\left(\frac{V_2}{V_1}\right)$$

where n is the amount of gas, R the gas constant, V_1 the initial volume, and V_2 the final volume.

Thus if 3 moles of an ideal gas expands from a volume of 5 dm^3 to a volume of 20 dm^3, we have

$$\begin{aligned}\Delta S &= 3\,\text{mol} \times 8.314\ \text{J K}^{-1}\ \text{mol}^{-1} \times \ln\left(\frac{20\,\text{dm}^3}{5\,\text{dm}^3}\right)\\ &= 24.942\ \text{J K}^{-1} \times \ln 4\\ &= 24.942\ \text{J K}^{-1} \times 1.386\\ &= 34.6\ \text{J K}^{-1}\end{aligned}$$

Questions

1. Write the following expressions in terms of logarithms:
   (a) $9 = 3^2$
   (b) $16 = 4^2$
   (c) $16 = 2^4$
   (d) $27 = 3^3$
   (e) $125 = 5^3$

2. Evaluate the following using a calculator:
   (a) $\ln 2.5$
   (b) $\ln 6.37$
   (c) $\ln 1.0$
   (d) $\ln 0.256$
   (e) $\ln 0.001$

3. Verify the following relationships using a calculator:
   (a) $\ln 20 = \ln 4 + \ln 5$
   (b) $\ln 10 = \ln 2 + \ln 5$
   (c) $\ln 10 = \ln 30 - \ln 3$
   (d) $\ln 6 = \ln 18 - \ln 3$
   (e) $\ln 9 = 2 \ln 3$

4. The entropy S of a system is related to the number W of possible configurations of a particular state of the system by $S = k \ln W$, where k is Boltzmann's constant. What is the value of S for a system of 3 molecules for which $W = 6$?

5. The change ΔG° in Gibbs free energy is related to the equilibrium constant K for a reaction by the equation $\Delta G^\circ = -RT \ln K$, where R is the gas constant, and T is the absolute temperature. Calculate ΔG° for the reaction

$$CH_3COOH_{(aq)} \rightarrow CH_3COO^-{}_{(aq)} + H^+{}_{(aq)}$$

   for which $K = 1.8 \times 10^{-5}$ at 298 K.

20. Logarithms to Base 10

The other logarithms commonly used in chemistry are those to base 10. Returning to the basic definition introduced in the previous chapter, i.e if

$$a = b^c \qquad \text{then } c = \log_b a$$

and considering simple examples such as

$$100 = 10^2 \qquad \text{so that } 2 = \log_{10} 100$$

and

$$1000 = 10^3 \qquad \text{so that } 3 = \log_{10} 1000$$

it becomes apparent that the logarithm to base 10 of a power of 10 is simply the power itself.

A little thought should suggest that representing a series of values with their logarithms gives a straightforward way of condensing the values of quantities which vary by several orders of magnitude.

The logarithm of a number x is normally abbreviated to $\log x$, i.e. if the subscript denoting the base is omitted then we assume that the base is 10.

Relationship between ln *x* and log *x*

The table below shows the values of $\ln x$ and $\log x$ for several values of x.

x	0.5	1.0	1.5	2.0	2.5	3.0
$\ln x$	−0.693	0.000	0.405	0.693	0.916	1.099
$\log x$	−0.301	0.000	0.176	0.301	0.398	0.477

In each case the ratio $\ln x/\log x$ has the value 2.303. It thus follows that:

$$\ln x = 2.303 \log x$$

Acidity of solution

This is usually measured using pH, which is defined as

$$\text{pH} = -\log_{10}([\text{H}^+]/\text{mol dm}^{-3})$$

Note the fact that the concentration $[H^+]$ is divided by its units. This ensures that we are taking the logarithm of a pure number; the result itself will also be a pure number without units.

The term p appears elsewhere in solution chemistry. For example, $pK_a = -\log 10\ K_a$, where K_a is the acidity constant.

Consider a solution of hydrochloric acid in which $[H^+] = 0.01$ mol dm^{-3}. Substituting into the definition above gives:

$$\text{pH} = -\log_{10}(0.01\ \text{mol dm}^{-3}/\text{mol dm}^{-3}) = -\log_{10}(0.01)$$

This is easily evaluated using a calculator, but writing 0.01 as 10^{-2} and simply taking the power -2 as being $\log_{10}10^{-2}$, we then have

$$\text{pH} = -(-2) = 2$$

which gives us a ready method for calculating the pH of an acidic solution if its hydrogen ion concentration can be written as an exact power of 10.

Beer–Lambert law

We saw in Chapter 15 that when light passes through a solution the absorbance A of the light is directly proportional to the concentration c of the solution. In fact, it is also proportional to the thickness l of solution through which the light passes and a constant ε known as the absorption coefficient. The Beer-Lambert law combines these relationships into one equation, $A = \varepsilon cl$.

The absorbance itself is defined in terms of the incident intensity, I_0, and the transmitted intensity, I, by the equation

$$A = \log_{10}\left(\frac{I_0}{I}\right)$$

Thus if $I = I_0$, then $I_0/I = 1$ and $A = \log_{10}1 = 0$, i.e. if the incident and transmitted beams have the same intensity there is zero absorbance.

The ratio of the transmitted to incident light intensities is defined as the transmittance, T, where

$$T = \frac{I}{I_0}$$

Consequently the reciprocal of T can be written as

$$T^{-1} = \frac{I_0}{I} \qquad \text{and so} \qquad A = \log T^{-1}$$

Using the rules of logarithms introduced in the previous chapter leads to

$$A = -\log T$$

Questions

1. Write the logarithm to base 10 of:
   (a) 10
   (b) 10^4
   (c) 10^8
   (d) 10^{-3}
   (e) 10^{-6}

2. Use the relationship between log and ln to determine:
   (a) $\ln 10^2$
   (b) $\ln 10^5$
   (c) $\ln 10^{10}$
   (d) $\ln 10^{-7}$
   (e) $\ln 0.01$

3. Evaluate the following using a calculator:
   (a) $\log 4.18$
   (b) $\log (3.16 \times 10^4)$
   (c) $\log (7.91 \times 10^{-4})$
   (d) $\log 0.003\,27$
   (e) $\log 3028$

4. What is the pH of an acidic solution having a hydrogen ion concentration of:
   (a) 0.01 mol dm^{-3}
   (b) 0.002 mol dm^{-3}
   (c) 5.0 mol dm^{-3}
   (d) 0.1014 mol dm^{-3}
   (e) 1.072 mol dm^{-3}

5. The Debye–Hückel limiting law gives the mean activity coefficient $\gamma_\pm$ of a pair of ions as:

$$\log \gamma_\pm = -(0.509\,\text{kg}^{1/2}\,\text{mol}^{-1/2})\,|z_+z_-|\,\sqrt{I}$$

   where z_+ and z_- are the units of charge on the respective ions, and I is the ionic strength. The term $|z_+z_-|$ means take the magnitude of this product with no regard to sign.

   Rewrite the Debye–Hückel limiting law for $\ln \gamma_\pm$.

6. The intensity of an incident beam is reduced by 60%. Calculate the absorbance and transmittance of this beam.

21. The Exponential Function

The exponential function is represented as

$$f(x) = e^x$$

where e takes the value 2.718. This is not an exact value, although it is programmed into electronic calculators far more precisely so normally this is not a problem.

To investigate the behaviour of this function it is useful to draw up a table of values, as below:

x	−2.5	−2.0	−1.5	−1.0	−0.5	0.0	0.5	1.0	1.5	2.0	2.5
$f(x)$	0.082	0.135	0.223	0.368	0.607	1.00	1.65	2.72	4.48	7.39	12.2

A graphical plot of these values is shown in Fig. 21.1.

First of all notice that the exponential function is never negative; the exponential of a negative number is always a fraction however. For positive values of x, the exponential function increases rapidly. This is what is meant by the commonly used term 'exponential growth', although strictly speaking this should only be used when the exact exponential function is used.

If you have verified any of the values in the above table, you may have noticed that to generate exponentials involved a sequence such as 'SHIFT' followed by 'ln'. There is a reason for this, which we will return to in Chapter 22.

The graph in Fig. 21.2 shows the behaviour of the function $f(x) = e^{-x}$, which is known as 'exponential decay'. Note that the value of the exponential becomes ever closer to zero but never actually reaches it.

First order kinetics

Kinetics is the study of the speeds with which reactions take place, and a study of it provides us with valuable information on the mechanisms of reactions. As we saw in Chapter 11, the rate of a reaction is quoted in units of mol dm^{-3} s^{-1} and shows how the concentration of a reactant varies with time. It is the mechanism of a reaction which determines its order n that appears in the general equation

$$rate = kc^n$$

where c is the concentration of reactant, and k a constant known as the rate constant which varies only with temperature. Thus for a first-order reaction, $n = 1$ and the equation becomes

$$rate = kc$$

Analysis of this equation, using techniques we have not yet met, leads to the relationship

$$c = c_0 e^{-kt}$$

which gives the concentration c of a reactant after time t in terms of its initial value c_0. A graph of c against t (Fig. 21.3) thus demonstrates this exponential decay.

The reaction

$$NH_2NO_{2(aq)} \rightarrow N_2O_{(g)} + H_2O_{(l)}$$

in alkaline solution is first order with a rate constant of 9.3×10^{-5} s^{-1}.

Consider a solution of NH_2NO_2 which is initially 0.15 mol dm^{-3}. To determine the concentration after 30 minutes, we first need to convert t into units consistent with k. Thus 30 minutes is equivalent to 30×60 s or 1800 s. We then simply substitute in the first-order rate equation.

$$\begin{aligned} c &= 0.15\,\text{mol dm}^{-3} \times \exp(-9.3 \times 10^{-5}\,\text{s}^{-1} \times 1800\,\text{s}) \\ &= 0.15\,\text{mol dm}^{-3} \times \exp(-0.1674) \\ &= 0.15\,\text{mol dm}^{-3} \times 0.846 \\ &= 0.13\,\text{mol dm}^{-3} \end{aligned}$$

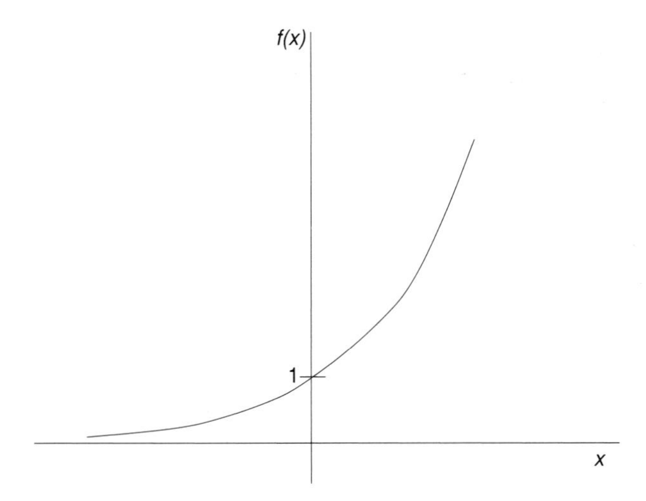

Figure 21.1 Graph of f(x) against x with $f(x) = e^x$

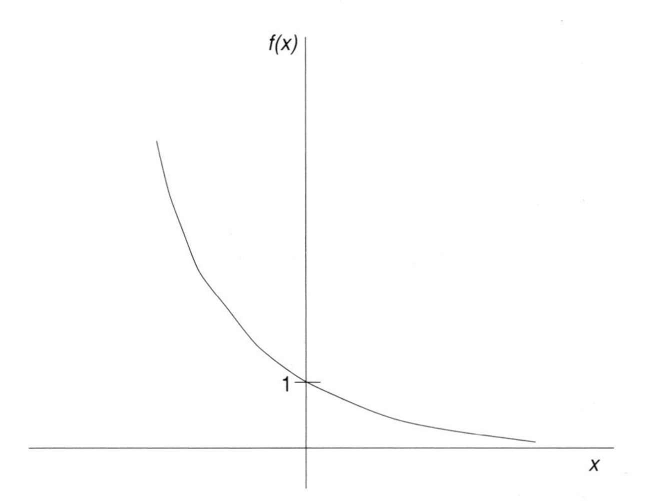

Figure 21.2 Graph of f(x) against x with $f(x) = e^{-x}$

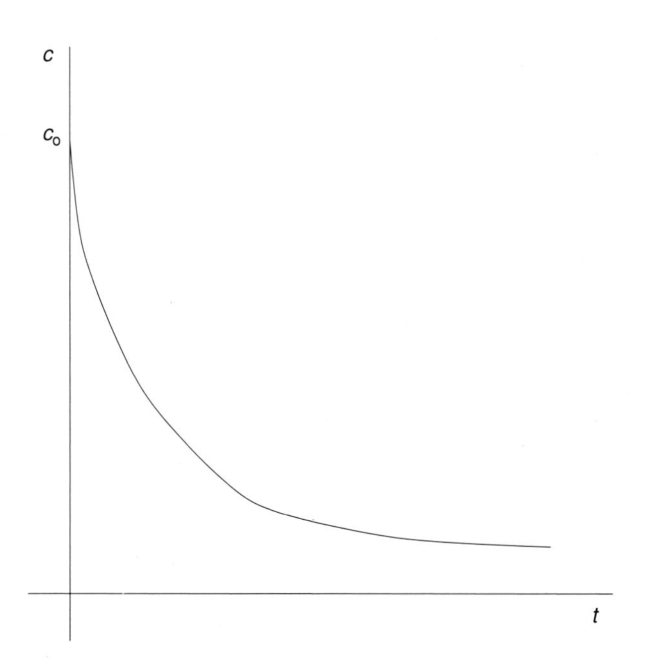

Figure 21.3 Graph of concentration c against time t for a first-order reaction

Notice that the units of s^{-1} and s within the exponential cancel. It is always the case that we can only take the exponential of a pure number without units. The result will also be a pure number without units. Note also that we have written $\exp(-0.1674)$ rather than $e^{-0.1674}$; this convention is often used for clarity, particularly when the expression within the exponential is complicated.

Boltzmann distribution law

The Boltzmann distribution law allows us to determine the fraction of molecules existing within two energy states ε_i and ε_j. If the number of molecules in each is n_i and n_j respectively, the ratio of these two populations is given by the equation

$$\frac{n_i}{n_j} = \exp\left[\frac{-(\varepsilon_i - \varepsilon_j)}{kT}\right]$$

where k is the Boltzmann constant (1.381×10^{-23} J K^{-1}), and T is the absolute temperature, expressed in Kelvin, K. Note the use of an expression of the form $\exp(x)$ rather than e^x for this relatively complicated expression.

The gas in a carbon dioxide laser consists of a mixture of helium, carbon dioxide and nitrogen. The rotational energy levels in carbon dioxide become populated through collisions with nitrogen, the former being 3.58×10^{-22} J higher in energy than those in the latter. The ratio of the number of carbon dioxide molecules to the number of nitrogen molecules with this populated level is given by

$$\begin{aligned}\frac{n_{CO_2}}{n_{N_2}} &= \exp\left[\frac{-3.58 \times 10^{-22}\,\text{J}}{1.38 \times 10^{-23}\,\text{J K}^{-1} \times 298\,\text{K}}\right] \\ &= \exp\left[\frac{-3.58 \times 10^{-22}\,\text{J}}{4.11 \times 10^{-21}\,\text{J}}\right] \\ &= \exp(-0.0871) \\ &= 0.917\end{aligned}$$

Note that this gives the ratio of the two populations. To obtain the fraction of molecules that have CO_2 populated, we have to divide by the total population, which is 0.917 + 1 or 1.917. The fraction required is then 0.917/1.917 or 0.478.

Questions

1. Evaluate:
   (a) e^2
   (b) e^{10}
   (c) $e^{1.73}$
   (d) $e^{2.65}$
   (e) $e^{9.9}$

2. Evaluate:
   (a) e^{-3}
   (b) e^{-7}
   (c) $e^{-2.19}$
   (d) $e^{-3.83}$
   (e) $e^{-4.7}$

3. The wavefunction Ψ for the 1s orbital of the hydrogen atom is given by

$$\Psi = \left(\frac{1}{\pi}\right)^{\frac{1}{2}}\left(\frac{1}{a_0}\right)^{\frac{3}{2}} e^{-r/a_0}$$

   where the Bohr radius $a_0 = 5.292 \times 10^{-11}$ m and $\pi = 3.142$. Calculate the value of Ψ when $r = 2.43 \times 10^{-11}$ m.

4. Radioactive decay is a first order process, so the number of nuclides n after time t is given by

$$n = n_0 e^{-kt}$$

   where n_0 is the initial number of nuclides present and k is the rate constant. The decay of uranium–238 has a rate constant k of 1.54×10^{-10} year^{-1}. What fraction of material has decayed after 4.51×10^9 years?

22. Inverse Functions

In simple terms an inverse function 'undoes' the effect of the original function. If the original function is $f(x)$, this inverse function is denoted by arc $f(x)$ or $f^{-1}(x)$. Note that the latter form $f^{-1}(x)$ does not mean the reciprocal of $f(x)$.

Suppose that a function $f(x)$ is defined as

$$f(x) = 4x + 3$$

It is helpful when thinking about inverse functions to consider what this means in words. This function could be described as 'take the input x, multiply by 4, and add 3'. To determine the inverse, we then undo each of these operations in reverse order. This would give 'subtract 3, and divide by 4'. In symbols, the definition is

$$\text{arc } f(x) = \frac{x-3}{4}$$

One of the particularly important cases of inverse functions used in chemistry is that involving logarithms and exponentials. The inverse of $\ln x$ is e^x, while the inverse of e^x is $\ln x$. Since the inverse undoes the effect of a function, it follows that

$$\ln e^x = x \quad \text{and} \quad e^{\ln x} = x$$

Similarly, for logarithms to base 10

$$\log 10^x = x \quad \text{and} \quad 10^{\log x} = x$$

so the inverse of $\log x$ is 10^x, and the inverse of 10^x is $\log x$.

The inverses of trigonometric functions are considered in Chapter 27.

The Arrhenius equation

This gives the variation of the rate constant k with absolute temperature T, and is expressed through the equation

$$k = Ae^{-E_a/RT}$$

where A is a constant known as the pre-exponential factor, E_a is the activation energy for the reaction, and R is the gas constant.

Taking natural logarithms of each side of this equation gives

$$\ln k = \ln (Ae^{\frac{-E_a}{RT}})$$

The right-hand side can be expanded using the rules of logarithms to give

$$\ln k = \ln A + \ln e^{\frac{-E_a}{RT}}$$

Since the final term involves taking the natural logarithm of an exponential, these functions effectively cancel each other out so that

$$\ln e^{\frac{-E_a}{RT}} = -\frac{E_a}{RT}$$

The transformed Arrhenius equation thus becomes

$$\ln k = \ln A - \frac{E_a}{RT}$$

Equilibrium constants

The standard change in the Gibbs free energy $\Delta G^{\ominus}$ for the reaction

$$\text{n-butane} \rightleftharpoons \text{isobutane}$$

is 6.6 kJ mol^{-1}. This value can be used to determine the equilibrium constant K for the reaction at 1000 K, using the relationship

$$\Delta G^{\ominus} = -RT \ln K$$

which was met in Chapter 19, where R is the gas constant, and T the absolute temperature.

Rearranging the equation gives

$$\ln K = -\frac{\Delta G^{\ominus}}{RT}$$

Substituting values into this gives

$$\ln K = -\frac{6.6 \times 10^3 \text{ J mol}^{-1}}{8.314 \text{ J K}^{-1} \text{ mol}^{-1} \times 1000 \text{ K}} = -0.794$$

If $\ln K = -0.794$, it follows that $K = e^{-0.794}$, which has the value 0.45. Note that we can only take the logarithm or exponential of a quantity without units; the result will also have no units. This can be a useful check in calculations.

Questions

1. Determine arc $f(x)$, if $f(x) = 4x + 7$.
2. Determine arc $f(x)$, if $f(x) = \ln(3x + 1)$.
3. Determine arc $g(y)$, if $g(y) = 2e^{3y}$.
4. In first-order reactions the concentration c varies with time t according to the equation

$$\ln c = \ln c_0 - kt$$

where c_0 is the initial concentration and k the rate constant. Obtain an expression for c without using logarithms.

5. The Nernst equation gives the electromotive force (EMF) E of an electrochemical cell as a function of the absolute temperature T as

$$E = E^{\ominus} - \frac{RT}{nF} \ln Q$$

where $E^{\ominus}$ is the EMF of the cell under standard conditions, R is the gas constant, n the amount of charge transferred, F the Faraday, and Q the reaction quotient. Determine the value of Q when $E = -0.029$ V, $E^{\ominus} = 0.021$ V, $R = 8.314$ J K^{-1} mol^{-1}, $T = 298$ K, $n = 2$, and $F = 96\,485$ C mol^{-1}.

6. The Debye–Hückel limiting law gives the activity coefficient $\gamma_{\pm}$ as

$$\log \gamma_{\pm} = -0.51\,|z_+ z_-| \sqrt{\left(\frac{I}{\text{mol dm}^{-3}}\right)}$$

where z_+ and z_- are the ionic charges (in units of e), and I is the ionic strength. Determine $\gamma_{\pm}$, if $z_+ = 2$, $z_- = 1$, and $I = 0.125$ mol dm^{-3}.

23. The Equation of a Straight Line

Straight line graphs are particularly important in chemistry. This is because they can be used to measure quantities and predict values in a more precise way than can be done for a curve.

Consider first of all the equation

$$y = 4x + 3$$

It is straightforward to construct a table showing different values of y for each chosen value of x. Such a table might look like:

x	0	1	2	3	4	5
y	3	7	11	15	19	23

If we plot these values we obtain a straight line (Fig. 23.1). The point where the line crosses the y-axis is called the intercept and in this case has the value 3. The gradient can be found by taking the distance between the y values, i.e. 23 – 3 or 20, and dividing by the corresponding difference in x values, i.e. 5 – 0 or 5. This gives a gradient of 20/5, or 4.

Comparison of these results with the equation above suggests that if we call the gradient m and the intercept c, the equation of a straight line can be written in general terms as $y = mx + c$.

Consider now the equation

$$y = 3x^2 - 1$$

Repeating the exercise above, the following table is obtained:

x	0	1	2	3	4	5
y	−1	2	11	26	47	74

Plotting these values gives a curve (Fig. 23.2). As noted above, this is much less straightforward to draw (at least by hand) and to obtain useful data from. It is much more useful if we can transform this data to give a straight line plot.

The table is repeated below, but an additional line has been added to give values of x^2.

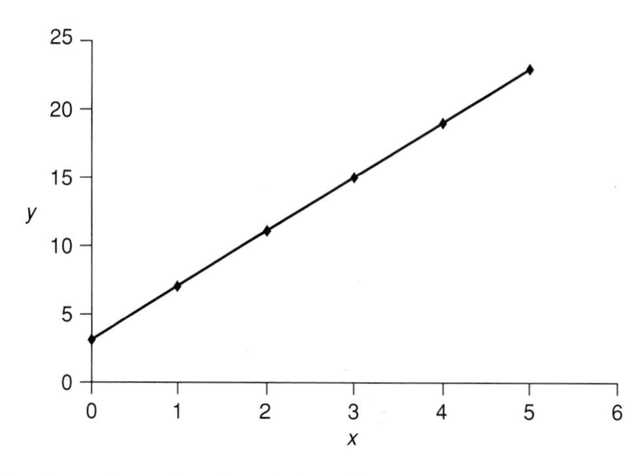

Figure 23.1 Graph of y against x for the relationship y = 4x + 3

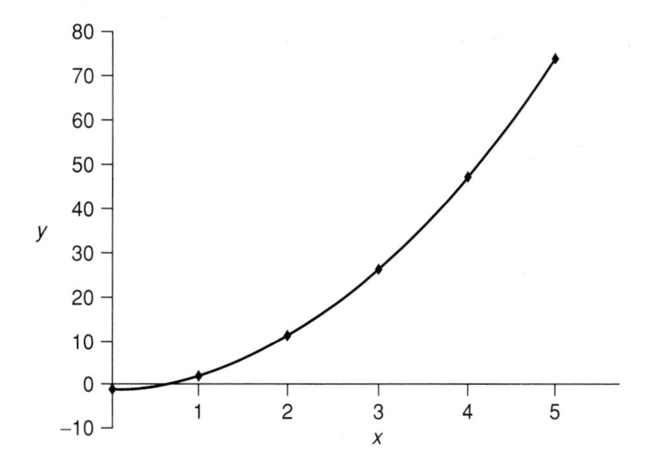

Figure 23.2 Graph of y against x for the relationship y = $3x^2 - 1$

x	0	1	2	3	4	5
y	−1	2	11	26	47	74
x^2	0	1	4	9	16	25

Plotting y against x^2, gives a straight line (Fig. 23.3). As you might expect from the previous example, the gradient is 3 and the intercept is −1.

This idea can be extended to include the y-axis. Thus a function of x called $X(x)$ and a function of y called $Y(y)$ are related by an equation of the form

$$Y = mX + c$$

A plot of $Y(y)$ against $X(x)$ will give a straight line having gradient m and intercept c.

Heat capacity of methane

The molar heat capacity C_p of methane varies with absolute temperature T according to the equation

$$\frac{C_p}{\text{J K}^{-1}\,\text{mol}^{-1}} = 22.34 + 0.0481\left(\frac{T}{\text{K}}\right)$$

If we rewrite this as

$$\frac{C_p}{\text{J K}^{-1}\,\text{mol}^{-1}} = 0.0481\left(\frac{T}{\text{K}}\right) + 22.34$$

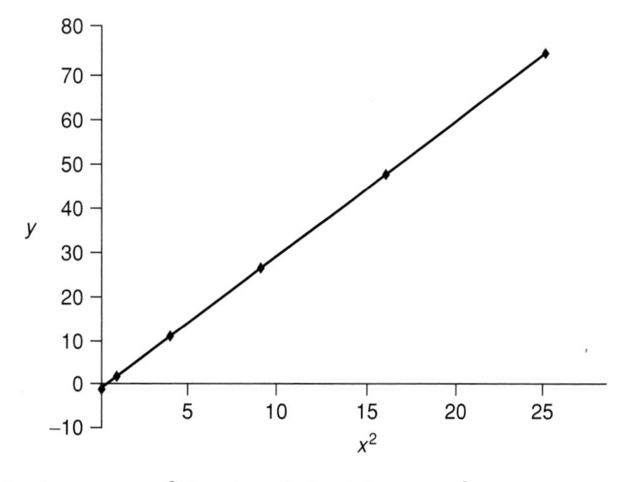

Figure 23.3 Graph of y against x^2 for the relationship y = $3x^2 - 1$

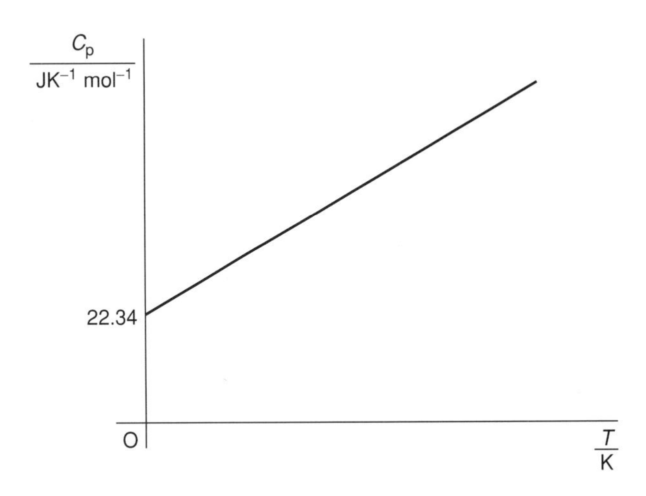

Figure 23.4 Graph of heat capacity C_p against absolute temperature T for methane

it is more obviously in the form $y = mx + c$. Comparing equivalent terms shows that if $C_p/\text{J K}^{-1}\ \text{mol}^{-1}$ is plotted on the y-axis against T/K on the x-axis, a straight line will be obtained having gradient 0.0481 and intercept 22.34. This can be verified by drawing up a table as before and plotting the graph (Fig. 23.4).

Beer–Lambert law

As seen previously in Chapters 15 and 20, this law is most commonly expressed in the form

$$A = \varepsilon c l$$

where A is the absorbance of light when it passes through a thickness l of a solution of concentration c having a molar absorption coeffiecient ε. Since l and ε are constant for a given experiment, A will be a function of c only, so grouping the constants in brackets allows us to write

$$A = (\varepsilon l)c$$

Comparison with the general equation of a straight line yields potential problems as the symbol for c is being used in two different contexts. In this case denote the intercept as k giving the general equation as $y = mx + k$. Comparison of equivalent terms then shows that $k = 0$ and the equation is actually of the form $y = mx$. Thus plotting A on the y-axis against c on the x-axis gives a straight line of gradient εl passing through the origin (Fig. 23.5).

Note that this can be used as the basis of a technique for determining ε, since l is generally known.

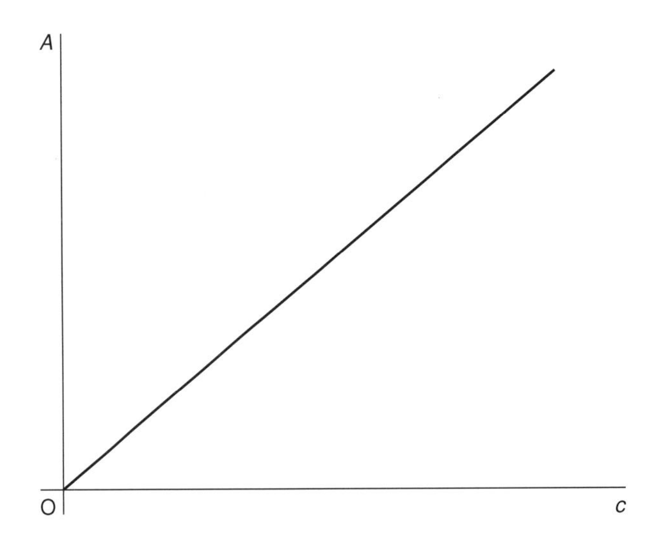

Figure 23.5 Beer–Lambert plot of absorbance A against concentration c

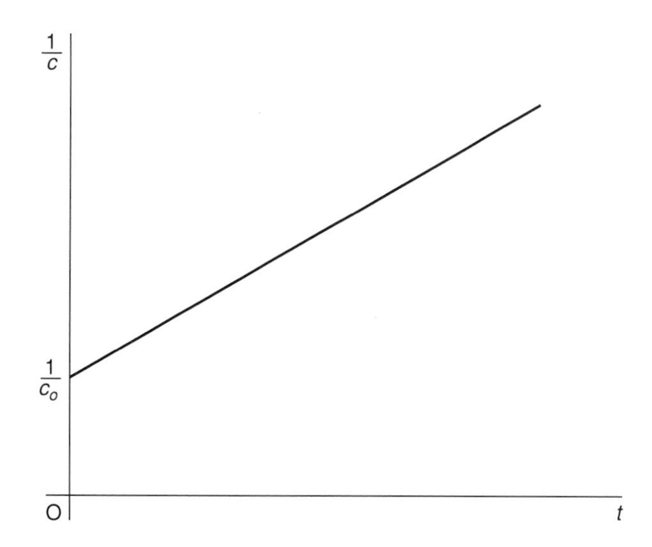

Figure 23.6 Graph of reciprocal of concentration 1/c against time t for a second-order reaction

Second-order kinetics

Some chemical reactions obey second-order kinetics, which means that their behaviour follows the equation

$$kt = \frac{1}{c} - \frac{1}{c_0}$$

where c is the concentration after time t, c_0 the initial concentration, and k the rate constant. This equation can be rearranged to give

$$\frac{1}{c} = kt + \frac{1}{c_0}$$

Comparison with $y = mx + c$ shows that a plot of $1/c$ on the y-axis against t on the x-axis will give a straight line with gradient k and intercept $1/c_0$ (Fig. 23.6).

Eadie–Hofstee plot

This is used when studying the kinetics of enzyme catalysis and is based on the Michaelis–Menten equation for the rate v of reaction

$$v = \frac{k_{cat}[E]_0}{1 + \frac{K_M}{[S]}}$$

where k_{cat} is known as the turnover number, $[E]_0$ is the initial enzyme concentration, K_M is the Michaelis constant, and $[S]$ is the concentration of substrate. In such an experiment typically values of the reaction rate v and the substrate concentration $[S]$ would be measured. Not surprisingly a plot of v against $[S]$ gives a curve, so the question arises as to how to determine the values of K_M and k_{cat} by constructing a linear plot.

To do this it is necessary to recast the formula which involves several steps. First divide both sides by $[E]_0$ to give

$$\frac{v}{[E]_0} = \frac{k_{cat}}{1 + \frac{K_M}{[S]}}$$

then multiply both sides by

$$1 + \frac{K_M}{[S]}$$

to give

$$\frac{v}{[E]_0}\left\{1 + \frac{K_M}{[S]}\right\} = k_{cat}$$

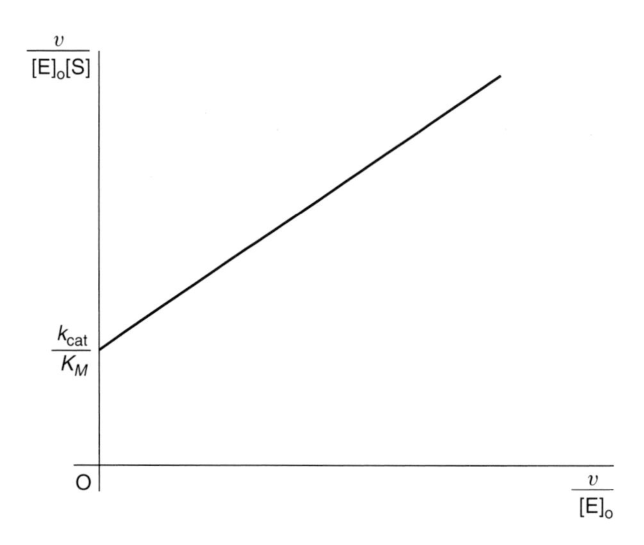

Figure 23.7 Graph of v/[E]0[S] against v/[E]0 for a reaction of a substrate S catalysed by an enzyme E obeying the Michaelis–Menten equation with rate v

Then convert the quantity within the curly brackets to a single fraction with denominator $[S]$:

$$\frac{v}{[E]_0}\left\{\frac{K_M + [S]}{[S]}\right\} = k_{cat}$$

Both sides of the equation are now divided by K_M to give

$$\frac{v}{[E]_0}\left\{\frac{K_M + [S]}{K_M[S]}\right\} = \frac{k_{cat}}{K_M}$$

Separating the quantity in curly brackets into two separate fractions leads to

$$\frac{v}{[E]_0}\left\{\frac{1}{[S]} + \frac{1}{K_M}\right\} = \frac{k_{cat}}{K_M}$$

Expanding the bracket (by multiplying each term within by that outside) gives

$$\frac{v}{[E]_0[S]} + \frac{v}{K_M[E]_0} = \frac{k_{cat}}{K_M}$$

Finally subtracting $\dfrac{v}{K_M[E]_0}$ from each side gives

$$\frac{v}{[E]_0[S]} = \frac{k_{cat}}{K_M} - \frac{v}{K_M[E]_0}$$

Comparing this with the general equation $y = mx + c$ shows that if $\frac{v}{[E]_0[S]}$ is plotted on the y-axis against $\frac{v}{[E]_0}$ on the x-axis, a straight line will be obtained with gradient $-1/K_M$ and intercept k_{cat}/K_M (Fig. 23.7).

Questions

1. If y is plotted against x for the following relationships, give the values of the resulting gradients and intercepts:
   (a) $y = 5x + 2$
   (b) $y = 3x - 7$
   (c) $2y = 4x - 9$
   (d) $x + y = 2$
   (e) $2x + 3y = 8$

2. For the following relationships, state what type of plot would give a straight line and the value of the resulting gradient and intercept:
   (a) $y = 3x^2 - 8$
   (b) $y^2 = 5x - 4$
   (c) $y = \frac{2}{x} - 3$

(d) $y^2 = \frac{3}{x} + 6$
(e) $y^2 = \frac{2}{x^2}$

3. For the following relationships, state what type of plot would give a straight line and the value of the resulting gradient and intercept:
(a) $x^2 + y^2 = 9$
(b) $2x^2 - y^2 = 5$
(c) $xy = 10$
(d) $x^2y = 4$
(e) $xy^2 - 14 = 0$

4. The integrated rate equations in kinetics give the concentration c of reactant after time t in terms of the rate constant k and initial concentration c_0. State what type of plot would give a straight line and the value of the resultant gradient and intercept for:
(a) zero-order kinetics with $c_0 - c = kt$
(b) second-order kinetics with $\left(\frac{1}{c} - \frac{1}{c_0}\right) = kt$

5. The vapour pressure p above a liquid is related to the absolute temperature T by the equation

$$\ln\left(\frac{p}{p^{\ominus}}\right) = -\frac{\Delta_{\text{vap}}H}{RT} + C$$

where $p^{\ominus}$ is the standard pressure of 1 atm, $\Delta_{\text{vap}}H$ is the enthalpy of vaporisation, R is the gas constant, and C is a constant. Explain how the value of $\Delta_{\text{vap}}H$ can be obtained from values of p and T using a graphical method.

6. The rate constant k varies with the absolute temperature T according to the Arrhenius equation, which as we saw in Chapter 22 can be expressed in the form

$$\ln k = \ln A - \frac{E_a}{RT}$$

where A is a constant, E_a is the activation energy, and R is the ideal gas constant. Explain how the value of E_a can be obtained from values of k and T using a graphical method.

24. Quadratic Equations

Quadratic equations are equations of the form

$$3x^2 - 5x + 2 = 0$$

which can be represented by the general expression

$$ax^2 + bx + c = 0$$

Note that although either or both b and c can be zero, the term in x^2 must be present in order for the equation to be quadratic.

The neatest way to solve a quadratic equation is, where possible, to use a process called factorisation. In the case of quadratic expressions, this involves rewriting the expression as a product of two terms. In the example above, this is

$$3x^2 - 5x + 2 = (3x - 2)(x - 1)$$

so that

$$(3x - 2)(x - 1) = 0$$

Since each bracket will represent a number once we have substituted for x, the product can only be zero if either or both of the brackets are zero. This can happen if $(3x - 2) = 0$, so that $3x = 2$ and $x = 2/3$, or if $(x - 1) = 0$, so that $x = 1$. The solutions of this equation are then $x = 2/3$ and $x = 1$. The fact that two solutions are obtained is a characteristic of a quadratic equation, although it is possible for both brackets to be equal giving the same solution twice.

One of the disadvantages of solving a quadratic equation in this way is that it takes skills to recognise the appropriate factorisation. More fundamentally, it is not always possible to factorise an expression.

The alternative form of solution is then to use a formula. If

$$ax^2 + bx + c = 0$$

then the solutions are given by

$$x = \frac{-b \pm \sqrt{(b^2 - 4ac)}}{2a}$$

Returning to our earlier example, and comparing

$$3x^2 - 5x + 2 = 0$$

with the more general form

$$ax^2 + bx + c = 0$$

gives $a = 3$, $b = -5$, and $c = 2$. These values can now be substituted into the formula for the general solution:

$$\begin{aligned} x &= \frac{-(-5) \pm \sqrt{\left[(-5)^2 - 4 \times 3 \times 2\right]}}{2 \times 3} \\ &= \frac{5 \pm \sqrt{(25 - 24)}}{6} \\ &= \frac{5 \pm \sqrt{1}}{6} \\ &= \frac{5 \pm 1}{6} \\ &= \frac{6}{6} \text{ or } \frac{4}{6} \end{aligned}$$

Dividing both top and bottom of the first of these by 6 clearly gives 1. The top and bottom of the second solution can be divided by 2 to give 2/3, so the same solutions are obtained as before, $x = 2/3$ and 1.

In chemistry generally the second method will be used, particularly as the coefficients a, b and c tend to have decimal values.

Dissociation of N_2O_4

One way of following the well-known equilibrium

$$N_2O_{4(g)} \rightleftharpoons 2\,NO_{2(g)}$$

is through the degree of dissocation α of the initial N_2O_4 present. It can be shown that this leads to the quadratic equation

$$4.054\,\alpha^2 + 0.1484\,\alpha - 0.1484 = 0$$

Comparing this with the general equation $ax^2 + bx + c = 0$, shows that $x = \alpha$, $a = 4.054$, $b = 0.1484$, and $c = -0.1484$. These values can then be substituted into the quadratic formula:

$$\begin{aligned}\alpha &= \frac{-0.1484 \pm \sqrt{\left[0.1484^2 - 4 \times 4.054 \times (-0.1484)\right]}}{2 \times 4.054} \\ &= \frac{-0.1484 \pm \sqrt{(0.0220 + 2.406)}}{8.108} \\ &= \frac{-0.1484 \pm \sqrt{2.428}}{8.108} \\ &= \frac{-0.1484 \pm 1.558}{8.108} \\ &= 0.174 \quad \text{or} \quad -0.210\end{aligned}$$

At this point we encounter a frequent problem in the solution of quadratic equations in chemistry, i.e. which solution is correct? A moment's thought suggest that the degree of dissociation must be positive, so the final answer is taken as $\alpha = 0.174$.

Pyrolysis of ethane

The overall equation for this reaction is

$$C_2H_6 \rightarrow C_2H_4 + H_2$$

but it actually takes place as a chain reaction with a series of steps. These are:

initiation	$C_2H_6 \rightarrow 2\,CH_3$	rate constant k_1
chain transfer	$CH_3 + C_2H_6 \rightarrow CH_4 + C_2H_5$	rate constant k_2
propagation	$C_2H_5 \rightarrow C_2H_4 + H$	rate constant k_3
	$H + C_2H_6 \rightarrow H_2 + C_2H_5$	rate constant k_4
termination	$H + C_2H_5 \rightarrow C_2H_6$	rate constant k_5

Such reaction schemes occur frequently in kinetics and can be analysed by considering the rates of reaction in which the intermediates CH_3, C_2H_5 and H are created and destroyed. In this case this results in the following set of equations:

$$2\,k_2[C_2H_6] - k_2[CH_3][C_2H_6] = 0$$

$$(k_2[CH_3] + k_4[H])[C_2H_6] - (k_3 + k_5[H])[C_2H_5] = 0$$

$$k_3[C_2H_5] - k_4[H][C_2H_6] - k_5[H][C_2H_5] = 0$$

which can be combined with some manipulation to give the quadratic equation

$$-2\,k_4k_5[H]^2 - 2\,k_1k_5[H] + 2\,k_1k_3 = 0$$

Although it is somewhat less obvious, this can be compared with the general equation to give $x =$ [H], $a = -2\,k_4k_5$, $b = -2\,k_1k_5$ and $c = 2\,k_1k_3$. Substituting into the quadratic formula gives:

$$[\mathrm{H}] = \frac{2\,k_1k_5 \pm \sqrt{\left[(-2\,k_1k_5)^2 - 4(-2\,k_4k_5)(2\,k_1k_3)\right]}}{2(-2\,k_4k_5)}$$

$$= \frac{2\,k_1k_5 \pm \sqrt{(4\,k_1^2k_5^2 + 16\,k_1k_3k_4k_5)}}{-4\,k_4k_5}$$

This is as far as this expression can be simplified in symbolic terms.

Questions

1. Solve the following equations using factorisation:
   (a) $x^2 + 3x - 10 = 0$
   (b) $x^2 - 3x = 0$
   (c) $3x^2 - 2x - 1 = 0$

2. Solve the following equations using the quadratic formula:
   (a) $2x^2 - 9x + 2 = 0$
   (b) $4x^2 + 4x + 1 = 0$
   (c) $3.6x^2 + 1.2x - 0.8 = 0$

3. For the dissociation of chlorine

$$Cl_2 \rightleftharpoons 2\,Cl$$

   the degree of dissociation is denoted by α. With an initial concentration of Cl_2 of 0.02 mol dm^{-3}, α is given by

$$0.04\,\alpha^2 + (1.715 \times 10^{-3})\,\alpha - (1.715 \times 10^{-3}) = 0$$

   Solve this equation to give the value of α.

4. In a 0.100 mol kg^{-1} solution of acetic acid the molality m of H_3O^+ is given by

$$m^2 + (1.75 \times 10^{-5})\,m - (1.75 \times 10^{-6}) = 0$$

   Solve this equation for m.

5. An additional 0.5 mol of NO_2 is added to the system

$$N_2O_{4(g)} \rightleftharpoons 2\,NO_{2(g)}$$

   so that the degree of dissociation α is now given by

$$4.0540\,\alpha^2 + 2.1750\,\alpha + 0.1054 = 0$$

   Determine the new value of α.

25. Sequences and Series

In Chapter 17 a function f of variable x, denoted by $f(x)$, was considered. If x can only take integral values, generally denoted as n, then $f(n)$ defines a sequence. For example, if

$$f(n) = 2n + 1 \quad \text{with } n \geq 1$$

then

$$f(1) = (2 \times 1) + 1 = 2 + 1 = 3$$
$$f(2) = (2 \times 2) + 1 = 4 + 1 = 5$$
$$f(3) = (2 \times 3) + 1 = 6 + 1 = 7$$

and so on.

The sum of the terms of a sequence, is said to define a series. To determine the sum of the terms above we could take

$$f(1)$$
$$f(1) + f(2)$$
$$f(1) + f(2) + f(3)$$

or any other number of terms. If the final term is defined by $f(N)$, then the sum of the full sequence will be

$$f(1) + f(2) + f(3) + \cdots + f(N)$$

This is cumbersome to write, so we use mathematical shorthand and express the sum as

$$\sum_{n=1}^{N} f(n)$$

Here Σ is the symbol for the Greek letter sigma, and simply means take the sum of whatever follows, in this case the general term $f(n)$. The $n = 1$ below Σ means start at this value while the N above denotes the final value of n.

There are certain series for which it is possible to determine precise formulae for their overall sums. One such example that is of interest to chemists is

$$1 + x + x^2 + x^3 + \cdots + x^{n-1}$$

for which the sum is

$$\frac{1 - x^n}{1 - x}$$

as long as $x \neq 1$.

First of all consider the first four terms of this expression if $x = 2$. This will give a sum of

$$1 + 2 + 2^2 + 2^3 = 1 + 2 + 4 + 8 = 15$$

Adding the next term x^4 adds 2^4 or 16; adding x^5 adds 2^5 or 32. Clearly as more terms are added the series becomes larger and larger. This series is said to diverge.

Now consider what happens if $x = 0.5$. The sums will be

$$1 + 0.5 + 0.5^2 + 0.5^3 = 1 + 0.5 + 0.25 + 0.125 = 1.875$$

The next term x^4 will add 0.5^4 or 0.0625; adding x^5 adds 0.5^5 or 0.031 25. This time consecutive terms added become smaller and smaller so that the series is said to converge to a limit.

The virial equation

The compressibility factor Z of a gas is defined by

$$Z = \frac{pV}{RT}$$

where p is its pressure, V the volume, T the absolute temperature, and R the gas constant. It is used as a measure of how far a real gas deviates from ideality. Given that the equation of state for an amount n of an ideal gas is

$$pV = nRT$$

it follows that $Z = 1$ denotes perfect ideality. In order to reproduce experimental data, Z can then be defined as a power series in V:

$$Z = 1 + \frac{B}{V} + \frac{C}{V^2} + \cdots$$

where B and C are constants, or less commonly as a power series in p:

$$Z = 1 + Bp + Cp^2$$

Note that the virial coefficients B and C are different in the two cases.

For the virial equation expressed in terms of volume, $B = -1.61 \times 10^{-5}\,\mathrm{m^3\,mol^{-1}}$ and $C = 1.2 \times 10^{-9}\,\mathrm{m^6\,mol^{-2}}$ for oxygen gas at room temperature.

Using the molar volume of an ideal gas, $25\,\mathrm{dm^3\,mol^{-1}}$, as an estimate of V, the relative contribution of each of these terms can be considered. First convert V to units of $\mathrm{m^3\,mol^{-1}}$, as follows:

$$\begin{aligned} V &= 25\,\mathrm{dm^3\,mol^{-1}} \\ &= 25 \times 10^{-3}\,\mathrm{m^3\,mol^{-1}} \\ &= 0.025\,\mathrm{m^3\,mol^{-1}} \end{aligned}$$

Then for oxygen

$$\frac{B}{V} = \frac{-1.61 \times 10^{-5}\,\mathrm{m^3\,mol^{-1}}}{0.025\,\mathrm{m^3\,mol^{-1}}} = -6.44 \times 10^{-4}$$

and

$$\begin{aligned} \frac{C}{V^2} &= \frac{1.2 \times 10^{-9}\,\mathrm{m^6\,mol^{-2}}}{(0.025\,\mathrm{m^3\,mol^{-1}})^2} \\ &= \frac{1.2 \times 10^{-9}\,\mathrm{m^6\,mol^{-2}}}{6.25 \times 10^{-4}\,\mathrm{m^6\,mol^{-2}}} \\ &= 1.92 \times 10^{-6} \end{aligned}$$

The numerical ratio of B/V to C/V^2 (ignoring the signs of the quantities because we are only interested in their sizes at this point) is therefore

$$\frac{6.44 \times 10^{-4}}{1.92 \times 10^{-6}} = 335$$

so that B/V makes a far greater contribution to Z. Note however that B/V and C/V^2 are both much smaller than 1 in this case so that Z remains close to 1.

Vibrational partition function

The subject of statistical mechanics was met briefly in Chapter 16. The underlying quantity which appears throughout the subject is that of the partition function q, which is defined for a molecule as the exponential function

$$q = \sum_i e^{-\varepsilon_i/kT}$$

where i are the integers which label the energy levels ε_i, k is Boltzmann's constant, and T the absolute temperature.

Since a molecule has translational, rotational, vibrational and electronic energy levels it is possible to define q for all four cases. Vibrational energy ε is given by $\varepsilon = h\nu$, where h is Planck's constant, and ν the frequency of vibration. The vibrational partition function q_v is then obtained by taking the sum over all the vibrational energy levels defined by the vibrational quantum number v. This is

$$\begin{aligned} q_v &= \sum_{v=0}^{\infty} e^{-vh\nu/kT} \\ &= e^0 + e^{-h\nu/kT} + e^{-2h\nu/kT} + e^{-3h\nu/kT} + \cdots \\ &= 1 + e^{-h\nu/kT} + (e^{-h\nu/kT})^2 + (e^{-h\nu/kT})^3 + \cdots \end{aligned}$$

since $e^{ax} = (e^x)^a$ from the rules of indices outlined in Chapter 1.

As mentioned earlier, the sum of $1 + x + x^2 + x^3 + \cdots + x^{n-1}$ is

$$\frac{1 - x^n}{1 - x}$$

so that in this case

$$q_v = \frac{1 - (e^{-h\nu/kT})^N}{1 - e^{-h\nu/kT}}$$

where N is a sufficiently large number, which we can use as an approximation to infinity.

Since

$$e^{-h\nu/kT} = \frac{1}{e^{h\nu/kT}}$$

and $e^{h\nu/kT}$ will be greater than 1, it follows that $(e^{-h\nu/kT})^N$ will be infinitesimal as N becomes very large. Consequently the overall expression for the partition function becomes

$$q_v = \frac{1}{1 - e^{-h\nu/kT}}$$

Questions

1. Determine whether the following series are convergent or divergent:
   (a) $1 + x + 2x^2 + 3x^3 + 4x^4 + \cdots$ for $x < 1$
   (b) $1 + \frac{1}{x!} + \frac{2}{(2x)!} + \frac{3}{(3x)!} + \frac{4}{(4x)!} + \cdots$ for $x > 1$
   (c) $1 + \ln x + 2\ln 2x + 3\ln 3x + 4\ln 4x + \cdots$ for $x < 1$

2. Determine the limits of the following series to 3 decimal places when $x = 4$:
   (a) $1 + \mathrm{e}^{-x} + \mathrm{e}^{-2x} + \mathrm{e}^{-3x} + \cdots$
   (b) $\frac{1}{x} + \frac{1}{2x^2} + \frac{1}{3x^3} + \frac{1}{4x^4} + \cdots$
   (c) $\frac{1}{\ln x} + \frac{1}{(\ln 2x)^2} + \frac{1}{(\ln 3x)^3} + \cdots$

3. For the virial equation expressed in terms of volume, $B = -4.5 \times 10^{-6}\ \mathrm{m^3\ mol^{-1}}$ and $C = 1.10 \times 10^{-9}\ \mathrm{m^6\ mol^{-2}}$ for nitrogen. Determine B/V, C/V^2 and their ratio.

4. The electronic partition function q_e is given by the sum

$$q_e = \sum_i g_i e^{-\varepsilon_i/kT}$$

   where g_i is the degeneracy of level i having energy ε_i, k being Boltzmann's constant, and T the absolute temperature. The ground state (level 0) of the oxygen atom has energy zero and degeneracy 5, level 1 has an energy of 3.15×10^{-21} J and degeneracy 3, and level 2 has energy 4.5×10^{-23} J and degeneracy 1. Calculate q_e for the oxygen atom at 298 K.

5. The rotational partition function q_r for a molecule is given by

$$q_r = \sum_{J=0}^{\infty} (2J + 1)e^{-J(J+1)h^2/8\pi^2 IkT}$$

   where J is the rotational quantum number, h is Planck's constant, I is the moment of inertia, k is Boltzmann's constant, and T is the absolute temperature. Write the first five terms of this series.

Spatial Mathematics

26. Trigonometry

The fundamental quantity in trigonometry is that of angular measurement. We are familiar with the fact that turning through a full circle takes us through 360°, and that 90° is a right angle. The unit of angular measurement is most commonly degrees.

There is however another way of measuring angles which is of great importance in science. This uses the unit radian, usually abbreviated to rad. The relationship between the two systems of units is expressed by:

$$\pi \text{ rad} = 180^\circ$$

where π is the well-known constant, which for many working purposes can be taken as the decimal number 3.142 or the fraction 22/7.

The above relationship allows us to determine the value of any angle in degrees or radians as appropriate. Since

$$\pi \text{ rad} = 180^\circ$$

dividing both sides by π gives

$$\frac{\pi \text{ rad}}{\pi} = \frac{180^\circ}{\pi}$$

or $1 \text{ rad} = \frac{180^\circ}{\pi}$

On the other hand, dividing the relationship

$$\pi \text{ rad} = 180^\circ$$

on both sides by 180 gives

$$\frac{\pi \text{ rad}}{180} = \frac{180^\circ}{180} \quad \text{or} \quad 1^\circ = (\pi/180) \text{ rad}$$

It is vital when working with a calculator to ensure that the correct mode is selected. It is usually possible to switch between DEG (degree) and RAD (radian) modes according to the problem being considered. Working in the wrong mode will give incorrect answers.

Trigonometric functions

Although many trigonometric functions are defined, we will only consider the three basic and most useful functions. These can all be defined in terms of the right-angled triangle shown in Fig. 26.1, where a stands for adjacent, o for opposite, and h for hypotenuse. These terms are defined relative to the angle θ.

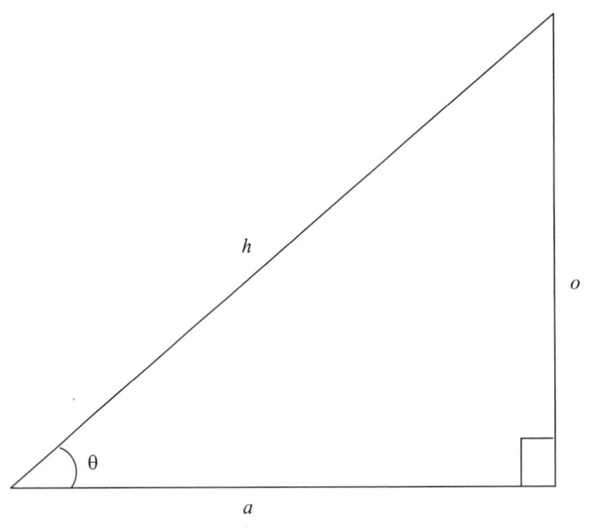

Figure 26.1 Right angled triangle used to define the trigonometric functions

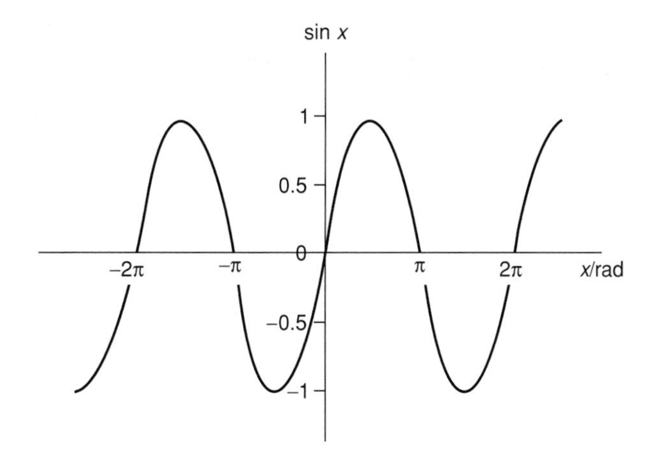

Figure 26.2 Graph of sin x against x

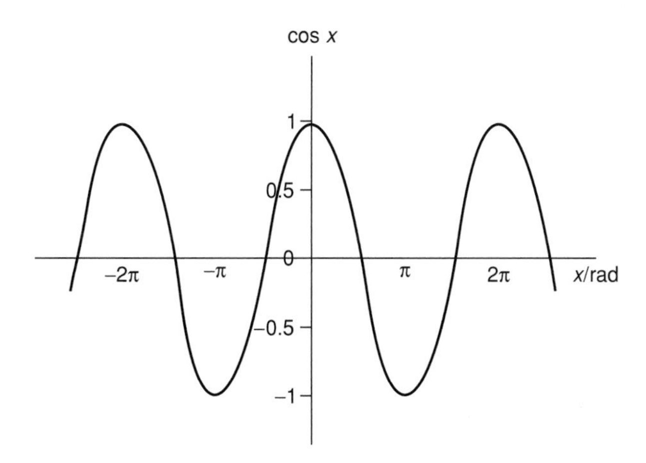

Figure 26.3 Graph of cos x against x

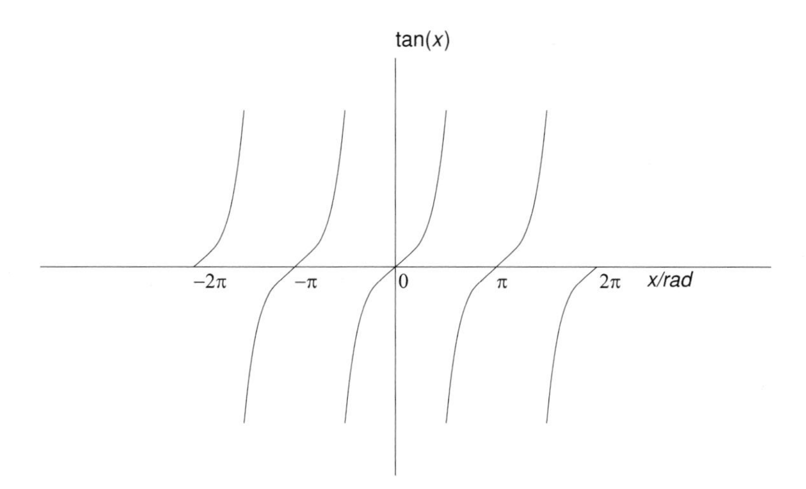

Figure 26.4 Graph of tan x against x

The three trigonometric functions sine, cosine and tangent are then defined by the equations:

$$\sin\,\theta = o/h \qquad \cos\,\theta = a/h \qquad \tan\,\theta = o/a$$

where the abbreviated form of each is used. The graphs of each of these functions are shown in Figs 26.2, 26.3 and 26.4, respectively. Note that while sine and cosine range from -1 to 1, tangent can have infinitely negative or positive values. Also note the periodic nature of the functions; this is to be expected as once one has turned through 360° the same values would be expected to occur. By convention positive angles refer to a clockwise turn with negative angles representing anticlockwise rotation.

Bragg's law

When X-rays strike a regular crystal they are diffracted. Bragg's law allows us to predict the angle θ at which such diffraction occurs, and is most commonly expressed as

$$n\lambda = 2d\sin\theta$$

where λ is the wavelength of radiation, and d is the spacing of crystal planes. The integer n is called the order or reflection; there will in fact be a number of values of θ for a given X-ray beam.

A set of crystal planes in the lithium fluoride crystal have a separation of 2.01×10^{-7} m. The first-order diffracted beam from these planes is seen at an angle θ of 34.7°. Since $n = 1$, the expression above becomes

$$1 \times \lambda = 2 \times 2.01 \times 10^{-7}\,\text{m} \times \sin 34.7^\circ$$

so

$$\begin{aligned}\lambda &= 2 \times 2.01 \times 10^{-7}\,\text{m} \times 0.569 \\ &= 2.29 \times 10^{-7}\,\text{m}\end{aligned}$$

Boundary conditions for particle in a one-dimensional box

Less obviously the trigonometric functions are important in quantum mechanics. The simple system of the particle in a one-dimensional box has been met previously in Chapters 13, 17 and 18; here the particle is able to move in the x-direction from $x = 0$ to $x = a$, a being the length of the box.

The particle will have an associated wavefunction, Ψ, which must have a value of 0 when $x = 0$ and when $x = a$; this is known as a boundary condition. The general form of the wavefunction $\Psi(x)$ is written as:

$$\Psi(x) = A\ \cos kx + B\ \sin kx$$

Such a combination of periodic functions is often used as a starting point when deriving quantum mechanical wavefunctions. A, B and k are constants whose values need to be determined.

It will be useful to refer to Figs 26.2 and 26.3, remembering that in this case we are working in radians.

First of all, when $x = 0$, we require that $\Psi(x) = 0$. Since $kx = 0$ and we see from the graph that $\sin 0 = 0$ and $\cos 0 = 1$, this means that we must have $A = 0$ in order to remove the cosine term.

When $x = a$, the wavefunction will be

$$\Psi(a) = B\sin ka$$

since we have already set $A = 0$. Looking at the graph shows that the sine function is zero when $x =$ 0 and also when x is a multiple of π. We must then have the condition

$$ka = n\pi$$

which can be rearranged to give

$$k = n\pi/a$$

Substituting back into our original equation now gives the wavefunction as

$$\Psi(x) = B\ \sin(n\pi x/a)$$

Further techniques are required in order to determine the value of B.

Questions

1. Evaluate the following:
   (a) $\sin 48^\circ$
   (b) $\cos 63^\circ$
   (c) $\tan 57^\circ$
   (d) $\sin(-32^\circ)$
   (e) $\cos 171^\circ$

2. Evaluate the following:
   (a) $\sin(\pi/3)$
   (b) $\cos(3/2)$

(c) tan 2
(d) $\sin(-\pi/7)$
(e) $\cos(5\pi/4)$

3. Use Bragg's law to determine the value of d in a crystal which diffracts X-rays of wavelength 154 pm at an angle of 12° for $n = 1$.

4. Sketch the first three wavefunctions ($n = 1, 2$, and 3) for the particle in a box.

5. If a point (x,y) is subjected to a rotation of θ about the z-axis, the resulting coordinates x' and y' are given by:

$$x' = x\cos\theta - y\sin\theta$$
$$y' = x\sin\theta + y\cos\theta$$

Determine the new coordinates of point (4,7) after a rotation of 35° about the z-axis.

6. The mean square end-to-end distance r^2 of a polymer chain is related to the angle θ between adjacent bonds by the equation:

$$r^2 = Nl^2\left(\frac{1 - \cos\theta}{1 + \cos\theta}\right)$$

where the chain consists of N bonds of length l. In a typical carbon atom chain, $l = 154$ pm, $N = 5 \times 10^3$, and $\theta = 109.5°$. Calculate the value of r.

27. Inverse Trigonometric Functions

In Chapter 22 it was seen that an inverse function 'undoes' the effect of a function. Also the exponential and natural logarithmic functions are the inverses of each other. In a similar fashion, each of the trigonometric functions considered in the previous chapter has an inverse.

$$\begin{aligned} &\text{The inverse of } \sin x \text{ is} \quad \sin^{-1} x \quad \text{or} \quad \arcsin x. \\ &\text{The inverse of } \cos x \text{ is} \quad \cos^{-1} x \quad \text{or} \quad \arccos x. \\ &\text{The inverse of } \tan x \text{ is} \quad \tan^{-1} x \quad \text{or} \quad \arctan x. \end{aligned}$$

It is very important to note that $\sin^{-1}$, $\cos^{-1}$, and $\tan^{-1}$ have nothing to do with the reciprocals of the trigonometric functions.

Suppose that $\tan \theta = 0.804$. The effect of the tan function is undone with $\tan^{-1}$, so that

$$\tan^{-1} (\tan \theta) = \tan^{-1} 0.804$$

which becomes

$$\theta = \tan^{-1} 0.804$$

This can be determined using a calculator and gives $\theta = 38.8°$ or 0.677 rad. Note that the inverse trigonometric functions on a calculator are usually accessed by pressing a combination of keys one of which includes the corresponding trigonometric function. Also, the answer will be displayed in degrees or radians, depending on the mode selected.

More complicated equations can be solved using the inverse trigonometric functions. Suppose that

$$\sin (3x + 2) = 0.762$$

As before, the appropriate inverse trigonometric function is used to give

$$\sin^{-1} \sin (3x + 2) = \sin^{-1} 0.762$$

which becomes

$$3x + 2 = \sin^{-1} 0.762$$

Using a calculator this becomes

$$3x + 2 = 0.866$$

Note that the answer is in radians; unless degrees are specified in the original equation, radians are taken as the default units.

Subtracting 2 from either side of this equation gives

$$3x + 2 - 2 = 0.866 - 2 \quad \text{or} \quad 3x = -1.134$$

then dividing each side by 3 gives

$$\frac{3x}{3} = \frac{-1.134}{3} \quad \text{or} \quad x = -0.378$$

Bragg's law

Bragg's law was introduced in the previous chapter as

$$n\lambda = 2d \sin \theta$$

where n is the order of diffraction, λ is the wavelength of X-rays, d is the spacing between crystal planes, and θ is the angle of diffraction. We are now interested in obtaining the angle of diffraction from the other data.

An X-ray tube with a copper target produces X-rays with a wavelength of 154 pm. In the iron sulphate crystal a set of planes known as (100) planes have a separation of 482 pm. Bragg's law then gives for the first order diffraction with $n = 1$

$$1 \times 154 \text{ pm} = 2 \times 482 \text{ pm} \times \sin\theta$$

so that

$$154 \text{ pm} = 964 \text{ pm} \times \sin\theta$$

Dividing both sides by 964 pm gives

$$\frac{154 \text{ pm}}{964 \text{ pm}} = \sin\theta \quad \text{or} \quad \sin\theta = 0.160$$

Notice that the units on the left-hand side of the equation cancel; the values of trigonometric functions and their inverses do not have units.

Now the appropriate inverse trigonometric function is used to give

$$\sin^{-1}\sin\theta = \sin^{-1} 0.160 \quad \text{or} \quad \theta = \sin^{-1} 0.160$$

Using a calculator set to degree mode now gives $\theta = 9.20°$.

Capillary action

If a narrow glass tube is inserted vertically into a liquid, the level of liquid rises in the tube. In some cases the liquid will not completely wet the glass, but will make an angle θ between its meniscus and the glass surface, as shown in Fig. 27.1. This angle is known as the contact angle. The surface tension γ, which is defined as the surface energy per unit area, is then given by the formula

$$\gamma = \frac{rh\rho g}{2\cos\theta}$$

where r is the internal radius of the tube, h is the height to which the liquid rises, ρ is the density of the liquid, and g is the acceleration due to gravity, a fixed constant having the value 9.81 m s^{-2}.

Mercury is an interesting example in that the level of liquid in the tube falls rather than rises; consequently the value of h is negative. The surface tension γ of mercury is 0.4355 N m^{-1} and its density is 13.53 g cm^{-3}. In a tube of radius 1 mm, the mercury level falls by 6.55 mm.

Converting each of these quantities to base units (Appendix 2) gives:

$$\gamma = 0.4355\,\text{N m}^{-1} = 0.4355\,\text{kg m s}^{-2}\,\text{m}^{-1} = 0.4355\,\text{kg s}^{-2}$$

$$r = 1\,\text{mm} = 10^{-3}\,\text{m}$$

$$h = -6.55\,\text{mm} = -6.55 \times 10^{-3}\,\text{m}$$

$$\rho = 13.53\,\text{g cm}^{-3} = 13.53 \times 10^{-3}\,\text{kg} \times (10^{-2}\,\text{m})^{-3} = 13.53 \times 10^{-3} \times 10^{6}\,\text{kg m}^{-3}$$
$$= 13.53 \times 10^{3}\,\text{kg m}^{-3}$$

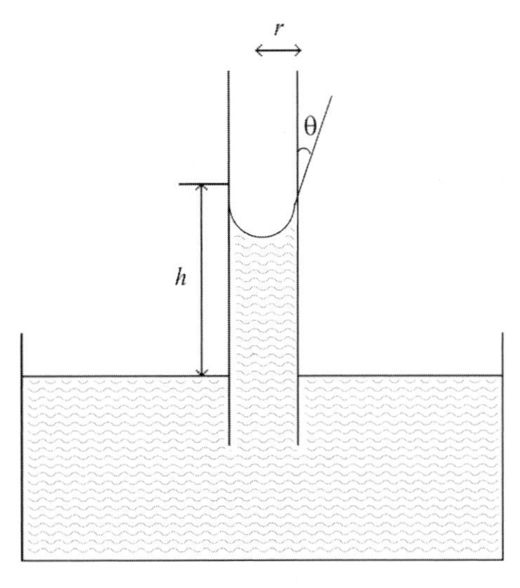

Figure 27.1 Angle θ between a liquid meniscus of height h and a capillary surface

The above equation for surface tension can be rearranged by multiplying each side by $\cos\theta$ to give

$$\gamma \cos\theta = \frac{rh\rho g}{2}$$

and then dividing each side by γ, so that

$$\cos\theta = \frac{rh\rho g}{2\gamma}$$

Substituting the values above now gives

$$\cos\theta = \frac{10^{-3}\,\mathrm{m} \times (-6.55 \times 10^{-3}\,\mathrm{m}) \times 13.53 \times 10^{3}\,\mathrm{kg\,m^{-3}} \times 9.81\,\mathrm{m\,s^{-2}}}{2 \times 0.4355\,\mathrm{kg\,s^{-2}}}$$

The units in this expression all cancel so that the numerical quantities can be combined to give

$$\cos\theta = -0.9981$$

The inverse trigonometric function is then used to give

$$\cos^{-1}\cos\theta = \cos^{-1}(-0.9981) = 176^{\circ}$$

Questions

1. Determine θ (in degrees) when:
   (a) $\sin\theta = 0.734$
   (b) $\cos\theta = -0.214$
   (c) $\tan\theta = 4.78$
   (d) $\sin\theta = -0.200$
   (e) $\tan\theta = -2.79$

2. Determine x (in radians) when:
   (a) $\sin x = 0.457$
   (b) $\cos x = 0.281$
   (c) $\tan x = 10.71$
   (d) $\sin x = -0.842$
   (e) $\cos x = -0.821$

3. Solve the following equations for x:
   (a) $\sin x - 0.815 = 0.104$
   (b) $\cos x + 0.421 = 0.817$
   (c) $\tan 3x = 5.929$
   (d) $\sin 2x - 0.318 = 0.520$
   (e) $\cos 4x + 0.212 = 0.957$

4. Bragg's law can be expressed in reciprocal space as

$$\sin\theta = \frac{\lambda}{2d_{hkl}}$$

   where d_{hkl} combines the distance between planes d and the order of reflection n, λ being the wavelength of incident radiation. Determine the value of θ when $\lambda = 132$ pm and $d_{hkl} = 220$ pm.

5. For a cubic crystal with unit cell dimension a

$$\sin^2\theta = \frac{\lambda^2}{4a^2}(h^2 + k^2 + l^2)$$

   for reflections from the planes denoted as $(h\,k\,l)$. Determine θ, if $\lambda = 136$ pm, $a = 313$ pm, $h = 1$, $k = 1$ and $l = 2$.

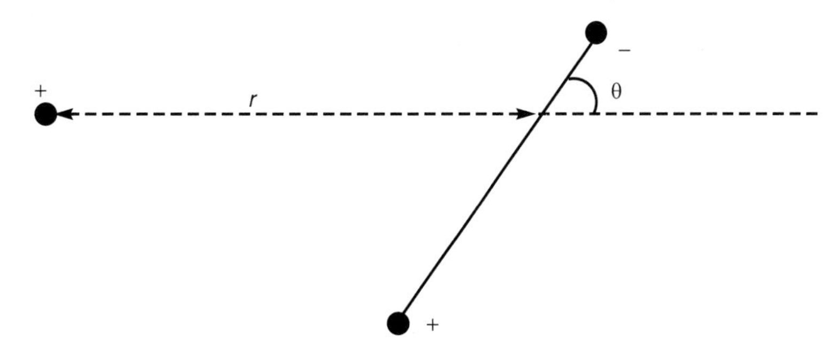

Figure 27.2 Interaction between an ion and a dipole

6. The interaction between an ion and a dipole is shown in Fig. 27.2. The energy E of interaction is given by

$$E = -\frac{z_A e \mu \cos\theta}{4\pi \varepsilon_0 \varepsilon_r r^2}$$

where z_A is the integer charge on the ion, e is the electronic charge, μ is the dipole moment, ε_0 is the permittivity of free space, and ε_r is the dielectric constant, which is also known as the relative permittivity. Determine θ if the ion–dipole interaction energy is –0.014 aJ, when $z_A = 2$, $e = 1.6 \times 10^{-19}$ C, $\mu = 2.50 \times 10^{-30}$ C m, $\varepsilon_0 = 8.85 \times 10^{-12}$ C^2 N^{-1} m^{-2}, $\varepsilon_r = 1.5$, and $r = 250$ pm.

28. Coordinate Geometry

A familiar example of coordinates in two dimensions is a map reference, which gives a unique identification of a point on the land. In order to set up such a coordinate system, an origin O needs to be defined, and two directions x and y. A general point P is then described as having coordinates (x,y).

If a line is drawn from O to P, and another parallel to the y direction, a right-angled triangle is formed, as shown in Fig. 28.1. The two shortest sides of the triangle have lengths x and y. It is then a simple matter to calculate the distance d from O to P using Pythagoras' theorem, which states that for a right-angled triangle such as this

$$d^2 = x^2 + y^2$$

so that

$$d = \sqrt{(x^2 + y^2)}$$

Suppose that there are now two points in this two-dimensional coordinate system. P has coordinates (x_1, y_1) while Q has coordinates (x_2, y_2), as shown in Fig. 28.2. A right-angled triangle can be constructed containing these points by drawing one line parallel to the x-axis and another parallel to the y-axis, as shown. The length of the horizontal line will be $(x_2 - x_1)$, while that of the vertical line will be $(y_2 - y_1)$. Using Pythagoras' theorem, the distance d between them will now be given by

$$d = \sqrt{\left[(x_2 - x_1)^2 + (y_2 - y_1)^2\right]}$$

Suppose there are two points, one at (1,4) and the other at (3,9). The distance d between them will be

$$\begin{aligned} d &= \sqrt{\left[(3-1)^2 + (9-4)^2\right]} \\ &= \sqrt{(2^2 + 5^2)} \\ &= \sqrt{(4 + 25)} \\ &= \sqrt{29} \\ &= 5.4 \end{aligned}$$

The values of x and y can both take negative values, in which case it can be very easy to deal with the negative signs incorrectly. Suppose the distance d between $(-2, -4)$ and $(-5, 8)$ is required. Using the formula gives

$$\begin{aligned} d &= \sqrt{\left[(-5-(-2))^2 + (8-(-4))^2\right]} \\ &= \sqrt{\left[(-5+2)^2 + (8+4)^2\right]} \\ &= \sqrt{\left[(-3)^2 + 12^2\right]} \\ &= \sqrt{(9 + 144)} \\ &= \sqrt{153} \\ &= 12.4 \end{aligned}$$

It is important to remember here that two negatives combine to make a positive, and that the square of a negative number is the same as the square of the corresponding positive number and is itself always positive.

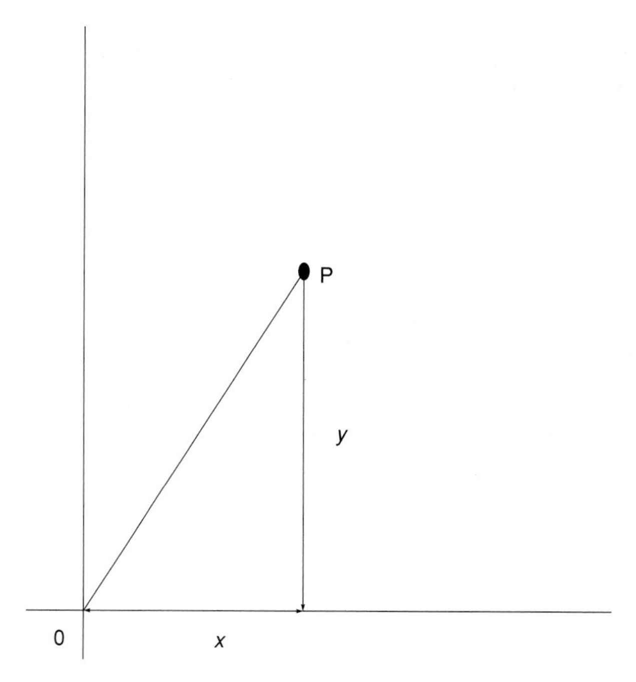

Figure 28.1 Coordinates of a point P (x, y) relative to the origin O

Coordinate systems in three dimensions

Most of the spatial problems encountered in chemistry are three dimensional, so a third direction, z, needs to be introduced to the previous coordinate scheme, as shown in Fig. 28.3. Calculating distances now requires inclusion of this third direction, so point P has coordinates (x_1, y_1, z_1) and point Q has coordinates (x_2, y_2, z_2). The distance d between them will be

$$d = \sqrt{\left[(x_2 - x_1)^2 + (y_2 - y_1)^2 + (z_2 - z_1)^2\right]}$$

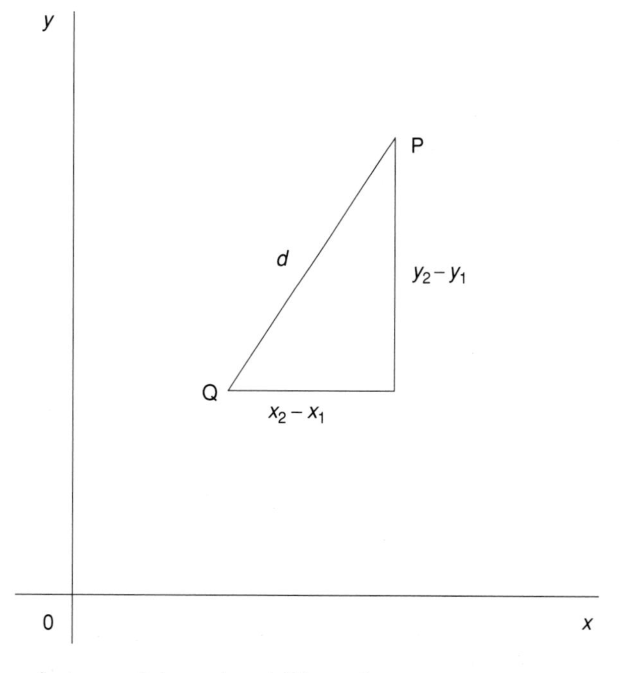

Figure 28.2 Distance between P (x_1, y_1) and Q(x_2, y_2)

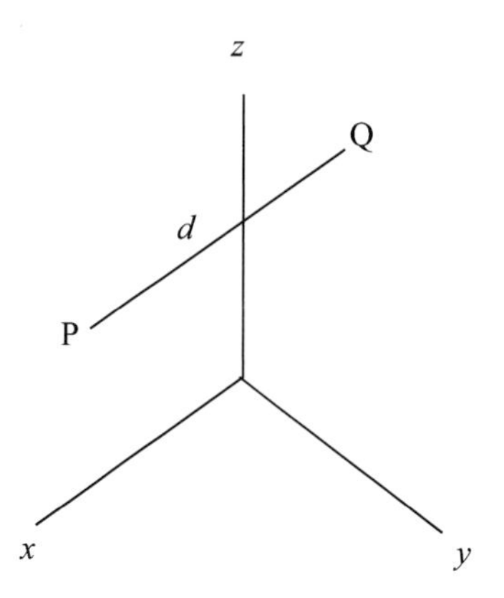

Figure 28.3 Distance between P(x_1, y_1, z_1) and Q(x_2, y_2, z_2) in three dimensions

If P is at (1, −2, −3) and Q at (4, 2, −1), then d will be given by

$$\begin{aligned} d &= \sqrt{\left[(4-1)^2 + (2-(-2))^2 + (-1-(-3))^2\right]} \\ &= \sqrt{(3^2 + 4^2 + 2^2)} \\ &= \sqrt{(9 + 16 + 4)} \\ &= \sqrt{29} \\ &= 5.4 \end{aligned}$$

Spherical polar coordinates

When a coordinate system is defined in terms of three axes which are mutually at right angles, the system is said to be orthogonal. It is however possible to define a coordinate system in other ways, although for every three-coordinate system three variables are required.

The spherical polar coordinate system uses three variables, r, θ, and ϕ, which are defined in Fig. 28.4. These are related to the orthogonal system by the series of equations

$$\begin{aligned} x &= r \sin\theta \cos\phi \\ y &= r \sin\theta \sin\phi \\ z &= r \cos\theta \\ r &= \sqrt{(x^2 + y^2 + z^2)} \\ \theta &= \cos^{-1}\left(\frac{z}{r}\right) \\ \phi &= \tan^{-1}\left(\frac{y}{x}\right) \end{aligned}$$

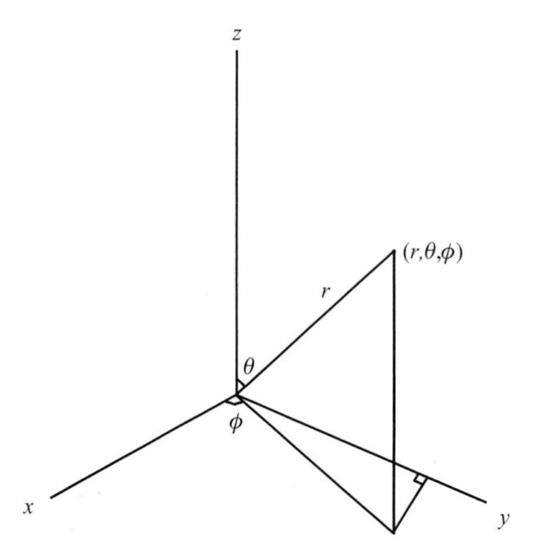

Figure 28.4 Definition of spherical polar coordinates

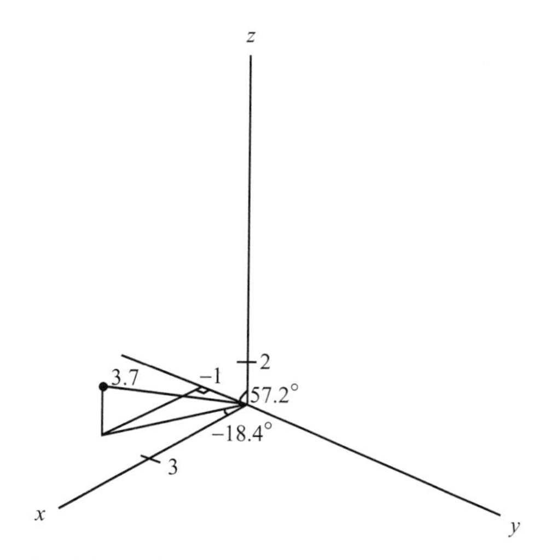

Figure 28.5 The point defined by orthogonal coordinates (3, −1, 2)

So for a point at (3, −1, 2) in the orthogonal system the corresponding spherical polar coordinates will be

$$\begin{aligned} r &= \sqrt{(3^2 + (-1)^2 + 2^2)} \\ &= \sqrt{(9 + 1 + 4)} \\ &= \sqrt{14} \\ &= 3.7 \\ \theta &= \cos^{-1}\left(\frac{2}{3.7}\right) \\ &= \cos^{-1} 0.541 \\ &= 57.2^\circ \\ \phi &= \tan^{-1}\left(\frac{-1}{3}\right) \\ &= \tan^{-1}(-0.333) \\ &= -18.4^\circ \end{aligned}$$

Note that the negative angle for ϕ represents an anticlockwise rotation, as shown in Fig. 28.5.

Now consider a point which in spherical coordinates is represented by (2.53, 47°, −10°), in other words $r = 2.53$, $\theta = 47^\circ$ and $\phi = -10^\circ$. Looking at the relationships given above shows that it is useful to evaluate

$$\begin{aligned} \sin\theta &= \sin 47^\circ = 0.7313 \\ \sin\phi &= \sin(-10^\circ) = -0.1736 \\ \cos\theta &= \cos 47^\circ = 0.6820 \\ \cos\phi &= \cos(-10^\circ) = 0.9848 \end{aligned}$$

Substituting into the appropriate relationships above gives

$$\begin{aligned} x &= r\sin\theta\cos\phi = 2.53 \times 0.7313 \times 0.9848 = 1.82 \\ y &= r\sin\theta\sin\phi = 2.53 \times 0.7313 \times (-0.1736) = -0.321 \\ z &= r\cos\theta = 2.53 \times 0.6820 = 1.73 \end{aligned}$$

These relationships are illustrated in Fig. 28.6.

Crystallography

When X-rays are fired at a regular crystal a diffraction pattern results (see Chapter 26). In single-crystal X-ray diffraction, careful analysis of the data allows us to determine the relative position of atoms in the repeat unit known as the unit cell. These positions can be specified in terms of x, y and z coordinates, and one coordinate system used is the orthogonal one described above. Coordinates are

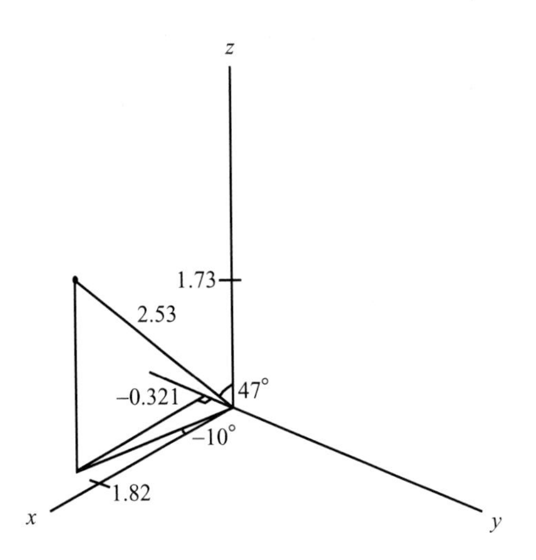

Figure 28.6 The point defined by spherical polar coordinates (2.53, 47°, −10°)

also frequently specified relevant to axes which are not mutually at right angles, but in these cases it is harder to calculate the distances between atoms.

The coordinates of eight of the carbon atoms in the diamond crystal structure are given in the table.

	x	y	z
C1	4.01625	0.44625	0.44625
C2	0.44625	0.44625	0.44625
C3	2.23125	2.23125	0.44625
C4	4.01625	4.01625	0.44625
C5	2.23125	0.44625	2.23125
C6	4.01625	2.23125	2.23125
C7	3.12375	1.33875	1.33875
C8	4.01625	0.44625	4.01625

The distance d between C1 and C7, for example, can be determined from the formula

$$\begin{aligned} d &= \sqrt{\left[(x_{C1} - x_{C7})^2 + (y_{C1} - y_{C7})^2 + (z_{C1} - z_{C7})^2\right]} \\ &= \sqrt{\left[(4.01625 - 3.12375)^2 + (0.44625 - 1.33875)^2 + (0.44625 - 1.33875)^2\right]} \\ &= \sqrt{\left[0.8925^2 + (-0.8925)^2 + (-0.8925)^2\right]} \\ &= \sqrt{(0.7966 + 0.7966 + 0.7966)} \\ &= \sqrt{2.3898} \\ &= 1.5459 \end{aligned}$$

This distance will take the units of the original coordinates. In this case these are Å, which is a non-SI unit equal to 10^{-10} m much used by crystallographers (Appendix 3). The distance d is therefore 1.5459 Å which is the standard length of a single bond between two carbon atoms. By calculating distances in this way crystallographers can determine which atoms in a crystal are bonded.

Quantum mechanics of the hydrogen atom

The hydrogen atom contains one electron which is free to move in space. The position of this electron can be described in three dimensions using three variables within a coordinate system of our choice.

One of the postulates of quantum mechanics is that the state of a particle, in this case an electron, is described fully by a wavefunction Ψ, from which all its observable properties can be determined. If an orthogonal coordinate system is being used, Ψ will be a function of x, y and z, and denoted more fully by $\Psi(x, y, z)$.

Intuitively an electron would not be expected to move parallel to the orthogonal x, y and z axes. In fact, it makes more sense to describe the wavefunction in terms of the spherical polar coordinates r, θ and ϕ. The first stage in analysing the hydrogen atom in quantum mechanics is therefore to transform $\Psi(x, y, z)$ into $\Psi(r, \theta, \phi)$. This is quite an involved process mathematically.

One advantage of doing this is that the three variables, r, θ and ϕ, are independent of each other, i.e. the value of one does not depend on that of another. Because of this it is possible to split $\Psi(r, \theta, \phi)$ into the product of three separate functions, each of which is a function of only one variable. Introducing the functions

R, which is a function only of the variable r, and is written as $R(r)$

Θ, which is a function only of the variable θ, and is written as $\Theta(\theta)$

Φ, which is a function only of the variable ϕ, and is written as $\Phi(\phi)$

These are chosen so that

$$\Psi(r, \theta, \phi) = R(r)\,\Theta(\theta)\,\Phi(\phi)$$

Equations can be set up for the three functions $R(r)$, $\Theta(\theta)$ and $\Phi(\phi)$, solved, and then combined to give the overall wavefunction $\Psi(r, \theta, \phi)$. This then allows various properties to be calculated, in particular the energy levels of the atom. Different energy levels have different associated wavefunctions; that for the ground state 1s level is

$$\Psi_{1s} = \frac{1}{\sqrt{\pi}}\left(\frac{1}{a_0}\right)^{3/2} e^{-r/a_0}$$

where a_0 is the Bohr radius 5.292×10^{-11} m, and r is the distance of the electron from the nucleus. Note that for the 1s energy level, the wavefunction Ψ is a function of only the variable r. For higher levels the wavefunction depends on the other variables; that for $2p_z$ is a function of both r and θ

$$\Psi_{2p_z} = \frac{1}{4\sqrt{2\pi}}\left(\frac{1}{a_0}\right)^{3/2}\left(\frac{r}{a_0}\right) e^{-r/2a_0}\cos\theta$$

while those for $2p_x$ and $2p_y$ depend on all three variables r, θ and ϕ

$$\Psi_{2p_x} = \frac{1}{4\sqrt{2\pi}}\left(\frac{1}{a_0}\right)^{3/2}\left(\frac{r}{a_0}\right) e^{-r/2a_0}\sin\theta\,\cos\phi$$

$$\Psi_{2_{p/y}} = \frac{1}{4\sqrt{2\pi}}\left(\frac{1}{a_0}\right)^{3/2}\left(\frac{r}{a_0}\right) e^{-r/2a_0}\sin\theta\,\sin\phi$$

Questions

1. Determine the distance between the following pairs of points:
   (a) (1, 2, 3) and (4, 0, 7)
   (b) (2, 0, 4) and (−4, 3, −2)
   (c) (8, −2, −5) and (7, −2, −7)

2. Convert the following orthogonal coordinates (x, y, z) to spherical polar coordinates (r, θ, ϕ):
   (a) (1, 2, 3)
   (b) (8, 7, 4)
   (c) (−1, 0, −9)

3. Convert the following spherical polar coordinates (r, θ, ϕ) to orthogonal coordinates (x, y, z):
   (a) $(6, \pi/2, \pi)$
   (b) $(10, -\pi/3, 2\pi)$
   (c) $(7.14, 35^\circ, -27^\circ)$

4. Determine which of the atoms C1–C8 in the diamond crystal structure are bonded to each other, i.e. are separated by a distance of approximately 1.54 Å.

5. For an electron in the 1s orbital the average value of r is $a_0/2$, where a_0 is the Bohr radius (Appendix 4). If $\theta = \phi = 45^\circ$, what are the cartesian coordinates of the electron?

6. Repeat question 5 for $r = a_0$, the most probable value of r, with $\theta = 60^\circ$ and $\phi = -30^\circ$.

29. Vectors

Most of the quantities we meet in chemistry are known as scalar quantities. Some, however, have directional properties and are known as vector quantities. To fully specify a vector quantity we need both its magnitude and direction.

Components of a vector

It is often useful to be able to split a vector into three components, which are usually in the x, y and z directions. The unit vectors in each of these directions are known as **i**, **j** and **k**, respectively; each is called a unit vector because it has magnitude 1. Notice that symbols for vector quantities are written using bold type.

Consider the vector $4\mathbf{i} + 2\mathbf{j} + 3\mathbf{k}$. To move from the start of the vector to its end we need to move 4 units in the x direction, 2 units in the y direction and 3 units in the z direction, as shown in Fig. 29.1. Geometrically the vector is represented by means of an arrow.

The vector $2\mathbf{i} - 3\mathbf{j} + \mathbf{k}$ would be formed by moving 2 units in the (positive) x direction, 3 units in the negative y direction, and 1 unit in the (positive) z direction.

Magnitude of a vector

From the coordinate geometry of the previous chapter the length of the vector in the first example above will be

$$\sqrt{}(4^2 + 2^2 + 3^2) = \sqrt{}(16 + 4 + 9) = \sqrt{}29 = 5.4$$

which is its magnitude. For the second example the magnitude of the vector is

$$\sqrt{}(2^2 + (-3)^2 + 1^2) = \sqrt{}(4 + 9 + 1) = \sqrt{}14 = 3.7$$

If a vector is denoted by **a**, then its magnitude is written as $|\mathbf{a}|$ or just a.

Addition of vectors

To add two vectors then simply add the individual components. So adding the two vectors above gives

$$(4\mathbf{i} + 2\mathbf{j} + 3\mathbf{k}) + (2\mathbf{i} - 3\mathbf{j} + \mathbf{k})$$

Then collecting together individual components

$$(4 + 2)\mathbf{i} + (2 - 3)\mathbf{j} + (3 + 1)\mathbf{k}$$

yields the overall vector

$$6\mathbf{i} - \mathbf{j} + 4\mathbf{k}$$

Geometrically the second vector is added to the first by ensuring that the arrows are in the same direction, as shown in Fig. 29.2.

Subtraction of vectors

Mechanically the subtraction of vectors is very similar to addition, with the individual components being combined. Subtracting the second vector above from the first gives

$$(4 - 2)\mathbf{i} + (2 - (-3))\mathbf{j} + (3 - 1)\mathbf{k}$$

which becomes

$$2\mathbf{i} + 5\mathbf{j} + 2\mathbf{k}$$

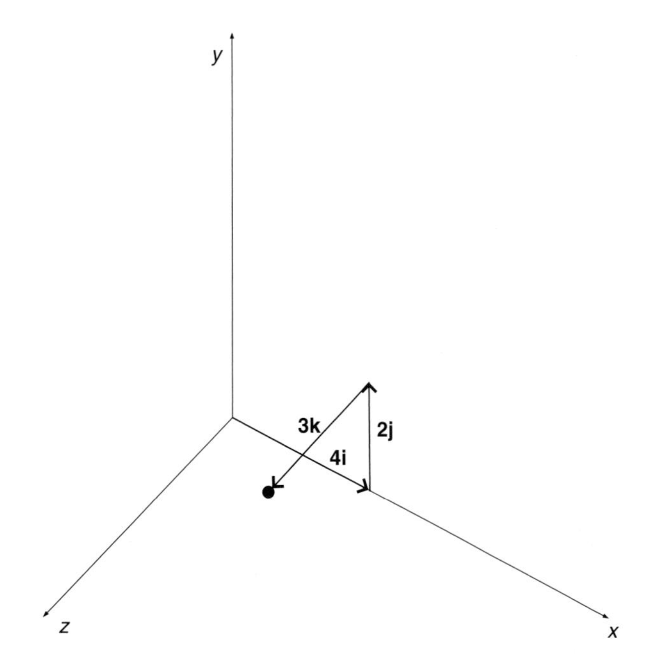

Figure 29.1 The vector 4**i** + 2**j** + **3k**

Geometrically this is represented by adding the reverse of the second vector to the first, as shown in Fig. 29.3. Effectively we have multiplied each component of the second vector by -1.

Determining a unit vector

There are occasions when a unit vector in a specified direction needs to be defined. This is done by dividing each component of the vector by the overall magnitude, so that for a vector **a** the unit vector $\hat{\mathbf{a}}$ in the same direction is given by

$$\hat{\mathbf{a}} = \frac{\mathbf{a}}{|\mathbf{a}|}$$

In the example where $\mathbf{a} = 4\mathbf{i} + 2\mathbf{j} + 3\mathbf{k}$ and $|\mathbf{a}| = 5.4$, we have

$$\hat{\mathbf{a}} = \frac{1}{5.4}(4\mathbf{i} + 2\mathbf{j} + 3\mathbf{k}) = 0.74\mathbf{i} + 0.37\mathbf{j} + 0.56\mathbf{k}$$

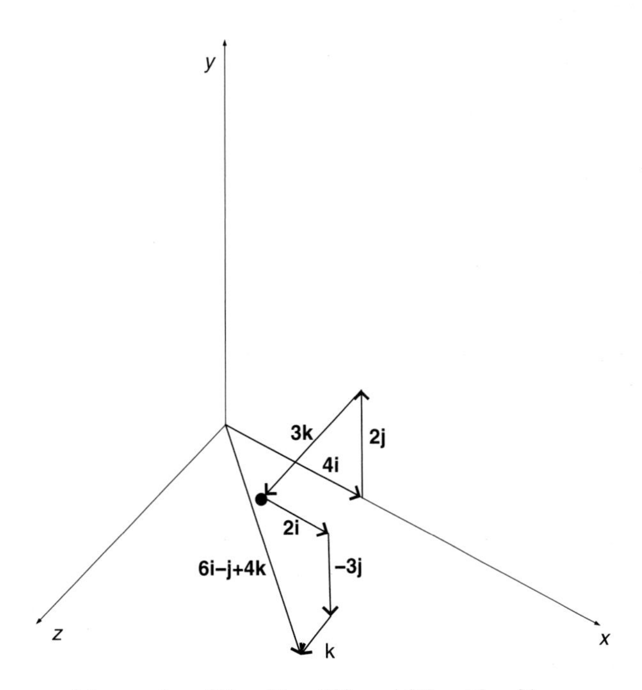

Figure 29.2 The sum of the vectors (4**i** + 2**j** + 3**k**) and (2**i** – 3**j** + **k**)

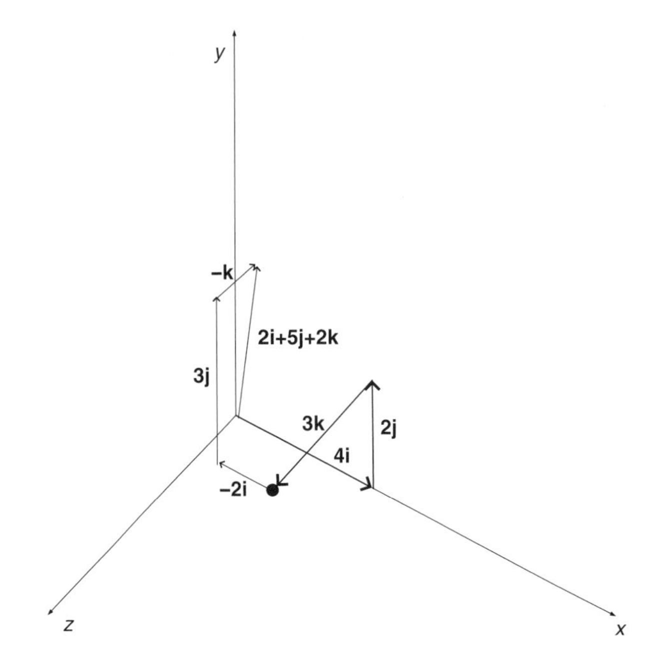

Figure 29.3 Difference between the vectors (4**i** + 2**j** + 3**k**) and (2**i** − 3**j** + **k**)

Coulomb's law

Coulomb's law defines the force F existing between two point charges Q_1 and Q_2. The magnitude of the force is given by the expression

$$F = \frac{1}{4\pi\varepsilon_o\varepsilon_r}\frac{Q_1Q_2}{r^2}$$

where ε_o is the permittivity of free space, and r the distance between the charges. The quantity ε_r was introduced in Chapter 27, and is known as either the dielectric constant or the relative permittivity. This is the form of Coulomb's law most commonly encountered but, since the force will act in a direction defined by a line between the charges, it should strictly be written as an equation involving vector quantities

$$\mathbf{F} = \frac{1}{4\pi\varepsilon_o\varepsilon_r}\frac{Q_1Q_2}{r^2}\hat{\mathbf{r}}$$

where $\hat{\mathbf{r}}$ is a unit vector along the line connecting the charges.

Angular momentum

An atom has both orbital angular momentum **L** and spin angular momentum **S**. These vector quantities can be added to give the total angular momentum **J**, i.e.

$$\mathbf{J} = \mathbf{L} + \mathbf{S}$$

The magnitudes of these quantities are written as J, L and S, which are known as quantum numbers for each of these quantities, and can be used to specify atomic states through the atomic term symbol

$$^{2S+1}L_J$$

The orbital angular momentum quantum number L is actually represented by a letter, which is S if $L = 0$, P if $L = 1$ and D if $L = 2$. Thus the atomic term symbol for helium in its ground state is $^1\mathrm{S}_0$, meaning that $2S + 1 = 1$ so $S = 0$, the symbol S indicating that $L = 0$ and the subscript $J = 0$.

Questions

1. Determine the magnitudes of the following vectors:
   (a) $3\mathbf{i} + 8\mathbf{j} + 9\mathbf{k}$
   (b) $5\mathbf{i} - 8\mathbf{j} + 8\mathbf{k}$
   (c) $-2\mathbf{i} + 9\mathbf{j} - 4\mathbf{k}$

2. Determine the sum $(\mathbf{a} + \mathbf{b})$ when:
   (a) $\mathbf{a} = 5\mathbf{i} + 6\mathbf{j} + 9\mathbf{k}$ and $\mathbf{b} = 3\mathbf{i} + 6\mathbf{j} + 2\mathbf{k}$
   (b) $\mathbf{a} = 2\mathbf{i} - 6\mathbf{j} + 9\mathbf{k}$ and $\mathbf{b} = 5\mathbf{i} - 8\mathbf{k}$
   (c) $\mathbf{a} = 9\mathbf{i} + 2\mathbf{j} - 2\mathbf{k}$ and $\mathbf{b} = 5\mathbf{i} - 3\mathbf{j} + 8\mathbf{k}$

3. Determine $(\mathbf{a} - \mathbf{b})$ for the vectors in question 2.

4. Determine the unit vector in the direction defined by:
   (a) $2\mathbf{i} + 6\mathbf{j} - 9\mathbf{k}$
   (b) $5\mathbf{i} - 8\mathbf{j} + 7\mathbf{k}$
   (c) $2\mathbf{i} + \mathbf{j} - 6\mathbf{k}$

5. Two points are defined by the positional vectors:

$$\mathbf{r}_1 = x_1\mathbf{i} + y_1\mathbf{j} + z_1\mathbf{k}$$
$$\mathbf{r}_2 = x_2\mathbf{i} + y_2\mathbf{j} + z_2\mathbf{k}$$

   Obtain an expression for the distance between the points by evaluating $|\mathbf{r}_2 - \mathbf{r}_1|$.

6. The translational vector $\mathbf{T}$ in a crystal is defined as

$$\mathbf{T} = n_1\mathbf{a} + n_2\mathbf{b} + n_3\mathbf{c}$$

   where $\mathbf{a}$, $\mathbf{b}$ and $\mathbf{c}$ are vectors which define the unit cell and n_1, n_2 and n_3 are integers. If $\mathbf{a}$ is in the x-direction, $\mathbf{b}$ in the y-direction and $\mathbf{c}$ in the z-direction, rewrite the expression for $\mathbf{T}$ in terms of the unit vectors $\mathbf{i}$, $\mathbf{j}$ and $\mathbf{k}$.

30. Vector Multiplication

We have already seen in the previous chapter that vectors may be added or subtracted. They may also be combined with each other or with scalar quantities by means of multiplication.

Multiplication by a scalar

The example of reversing a vector by multiplying by -1 was met in the previous chapter. In fact, a vector can be multiplied by any scalar number. For example, if the vector $\mathbf{i} - \mathbf{j} + 2\mathbf{k}$ is multiplied by 3, we have

$$3(\mathbf{i} - \mathbf{j} + 2\mathbf{k})$$

which multiplying term by term gives

$$3\mathbf{i} - 3\mathbf{j} + 6\mathbf{k}$$

In terms of symbols, if a vector $\mathbf{v}$ is multiplied by a scalar a, the result is written as $a\mathbf{v}$. This technique can also be used to divide a vector by a scalar. This is often done by writing the dividing number as its reciprocal, which then multiplies the vector. If $2\mathbf{i} + 2\mathbf{j} - 4\mathbf{k}$ is divided by 2, we have

$$\frac{1}{2}(2\mathbf{i} + 2\mathbf{j} - 4\mathbf{k})$$

which becomes $\mathbf{i} + \mathbf{j} - 2\mathbf{k}$. In symbols, if vector $\mathbf{v}$ is divided by a, this would be written as

$$\frac{\mathbf{v}}{a} \quad \text{or} \quad \frac{1}{a}\mathbf{v}$$

This was done in the previous chapter when determining the magnitude of a vector.

Scalar product

As the name implies, the scalar product is a way of multiplying vectors which results in a scalar quantity. It is also known as the 'dot product,' because the multiplication operation is represented by a dot. The scalar product of two vectors $\mathbf{a}$ and $\mathbf{b}$ is defined by

$$\mathbf{a} \cdot \mathbf{b} = |\mathbf{a}||\mathbf{b}| \cos\theta$$

where $|\mathbf{a}|$ and $|\mathbf{b}|$ are the magnitudes of $\mathbf{a}$ and $\mathbf{b}$ respectively and θ is the angle between them. Notice that the angle θ is defined as in Fig. 30.1, where both vectors point away from their intersection point.

If we consider the behaviour of the unit vectors $\mathbf{i}$, $\mathbf{j}$ and $\mathbf{k}$, which each have magnitude 1, we obtain the relationships

$$\mathbf{i} \cdot \mathbf{i} = \mathbf{j} \cdot \mathbf{j} = \mathbf{k} \cdot \mathbf{k} = 1$$

since in all three cases the vectors being combined are in the same direction with $\theta = 0^\circ$ and $\cos\theta = 1$. Similarly

$$\mathbf{i} \cdot \mathbf{j} = \mathbf{i} \cdot \mathbf{k} = \mathbf{j} \cdot \mathbf{k} = 0$$

since in all three cases the vectors being combined are perpendicular with $\theta = 90^\circ$ and $\cos\theta = 0$. Notice also that $\mathbf{a} \cdot \mathbf{b} = \mathbf{b} \cdot \mathbf{a}$, for any pair of vectors.

Using these relationships we can now combine any pair of vectors that are expressed in terms of the unit vectors $\mathbf{i}$, $\mathbf{j}$ and $\mathbf{k}$. Consider the vectors

$$\mathbf{a} = 2\mathbf{i} + \mathbf{j} - \mathbf{k} \quad \text{and} \quad \mathbf{b} = \mathbf{i} - 2\mathbf{j} + \mathbf{k}$$

To evaluate $\mathbf{a} \cdot \mathbf{b}$, we need to consider every product of unit vectors within

$$(2\mathbf{i} + \mathbf{j} - \mathbf{k}) \cdot (\mathbf{i} - 2\mathbf{j} + \mathbf{k})$$

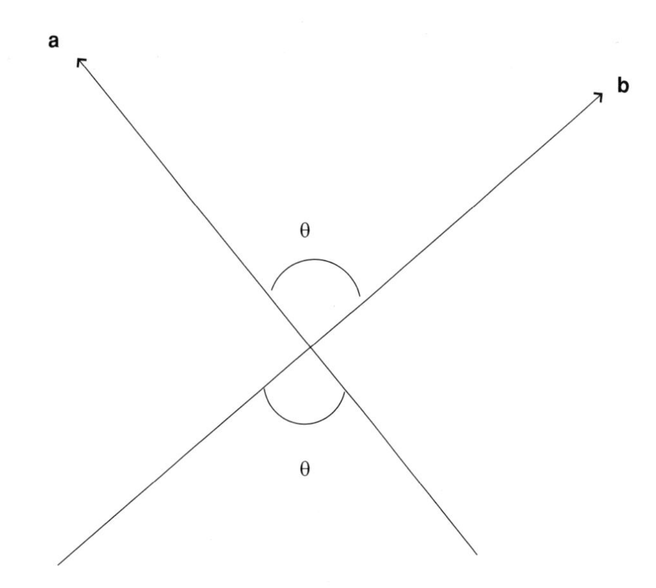

Figure 30.1 Definition of the angle θ between two vectors

This gives

$$\begin{aligned}
&2\mathbf{i}\cdot\mathbf{i}+2\mathbf{i}\cdot(-2\mathbf{j})+2\mathbf{i}\cdot\mathbf{k}+\mathbf{j}\cdot\mathbf{i}+\mathbf{j}\cdot(-2\mathbf{j})+\mathbf{j}\cdot\mathbf{k}-\mathbf{k}\cdot\mathbf{i}-\mathbf{k}\cdot(-2\mathbf{j})-\mathbf{k}\cdot\mathbf{k}\\
&\quad=2\mathbf{i}\cdot\mathbf{i}-4\mathbf{i}\cdot\mathbf{j}+2\mathbf{i}\cdot\mathbf{k}+\mathbf{j}\cdot\mathbf{i}-2\mathbf{j}\cdot\mathbf{j}+\mathbf{j}\cdot\mathbf{k}-\mathbf{k}\cdot\mathbf{i}+2\mathbf{k}\cdot\mathbf{j}-\mathbf{k}\cdot\mathbf{k}\\
&\quad=2\times1-4\times0+2\times0+0-2\times1+0-0+2\times0-1\\
&\quad=2-2-1=-1
\end{aligned}$$

Vector product

The vectors **a** and **b** may also be combined to give a vector. The vector product, or cross product, is defined by

$$\mathbf{a}\times\mathbf{b}=|\mathbf{a}||\mathbf{b}|\sin\theta\,\hat{\mathbf{n}}$$

where the symbols have the same meaning as before, and $\hat{\mathbf{n}}$ is a unit vector perpendicular to both **a** and **b** in a direction given by a right-handed clockwise turn from **a** to **b**.

This time for the unit vectors **i**, **j** and **k**, we have

$$\mathbf{i}\times\mathbf{j}=\mathbf{k}\qquad\mathbf{j}\times\mathbf{k}=\mathbf{i}\qquad\mathbf{k}\times\mathbf{i}=\mathbf{j}$$

since $\theta = 90^\circ$ and $\sin\theta = 1$. However, reversing the order of the vectors reverses the direction of $\hat{\mathbf{n}}$ and so

$$\mathbf{j}\times\mathbf{i}=-\mathbf{k}\qquad\mathbf{k}\times\mathbf{j}=-\mathbf{i}\qquad\mathbf{i}\times\mathbf{k}=-\mathbf{j}$$

When the unit vectors in the vector product are the same we have

$$\mathbf{i}\times\mathbf{i}=\mathbf{j}\times\mathbf{j}=\mathbf{k}\times\mathbf{k}=\mathbf{0}$$

since $\theta = 0^\circ$ and $\sin\theta = 0$.

The vector product can now be determined for any pair of vectors expressed in terms of the unit vectors **i**, **j** and **k**. If

$$\mathbf{c}=2\mathbf{i}-\mathbf{j}+3\mathbf{k}\qquad\text{and}\qquad\mathbf{d}=\mathbf{i}-2\mathbf{j}+2\mathbf{k}$$

Then the vector product is

$$\begin{aligned}
\mathbf{c}\times\mathbf{d}&=(2\mathbf{i}-\mathbf{j}+3\mathbf{k})\times(\mathbf{i}-2\mathbf{j}+2\mathbf{k})\\
&=2\mathbf{i}\times\mathbf{i}+2\mathbf{i}\times(-2\mathbf{j})+2\mathbf{i}\times2\mathbf{k}-\mathbf{j}\times\mathbf{i}-\mathbf{j}\times(-2\mathbf{j})-\mathbf{j}\times2\mathbf{k}+3\mathbf{k}\times\mathbf{i}+3\mathbf{k}\\
&\quad\times(-2\mathbf{j})+3\mathbf{k}\times2\mathbf{k}\\
&=2\mathbf{i}\times\mathbf{i}-4\mathbf{i}\times\mathbf{j}+4\mathbf{i}\times\mathbf{k}-\mathbf{j}\times\mathbf{i}+2\mathbf{j}\times\mathbf{j}-2\mathbf{j}\times\mathbf{k}+3\mathbf{k}\times\mathbf{i}-6\mathbf{k}\times\mathbf{j}+6\mathbf{k}\times\mathbf{k}\\
&=2\times0-4\mathbf{k}+4(-\mathbf{j})-(-\mathbf{k})+2\times0-2\mathbf{i}+3\mathbf{j}-6(-\mathbf{i})+6\times0\\
&=-4\mathbf{k}-4\mathbf{j}+\mathbf{k}-2\mathbf{i}+3\mathbf{j}+6\mathbf{i}=4\mathbf{i}-\mathbf{j}-3\mathbf{k}
\end{aligned}$$

Angular momentum

The classical angular momentum **L** is defined by the vector product

$$\mathbf{L} = \mathbf{r} \times \mathbf{p}$$

where **r** is the distance from the origin of a particle of mass m rotating about a fixed point with velocity **v** having linear momentum **p** which is equal to $m\mathbf{v}$. The distance **r** can be described by means of a position vector in terms of the particle's position (x, y, z) and the unit vectors **i**, **j** and **k**

$$\mathbf{r} = x\mathbf{i} + y\mathbf{j} + z\mathbf{k}$$

The momentum **p** can be partitioned into components in the x, y and z directions denoted as p_x, p_y and p_z respectively so that

$$\mathbf{p} = p_x\mathbf{i} + p_y\mathbf{j} + p_z\mathbf{k}$$

The vector product is now

$$\begin{aligned}
\mathbf{L} &= (x\mathbf{i} + y\mathbf{j} + z\mathbf{k}) \times (p_x\mathbf{i} + p_y\mathbf{j} + p_z\mathbf{k}) \\
&= x\mathbf{i} \times p_x\mathbf{i} + x\mathbf{i} \times p_y\mathbf{j} + x\mathbf{i} \times p_z\mathbf{k} + y\mathbf{j} \times p_x\mathbf{i} + y\mathbf{j} \times p_y\mathbf{j} + y\mathbf{j} \times p_z\mathbf{k} + z\mathbf{k} \times p_x\mathbf{i} + z\mathbf{k} \times p_y\mathbf{j} \\
&\quad + z\mathbf{k} \times p_z\mathbf{k} \\
&= xp_x\mathbf{i} \times \mathbf{i} + xp_y\mathbf{i} \times \mathbf{j} + xp_z\mathbf{i} \times \mathbf{k} + yp_x\mathbf{j} \times \mathbf{i} + yp_y\mathbf{j} \times \mathbf{j} + yp_z\mathbf{j} \times \mathbf{k} + zp_x\mathbf{k} \times \mathbf{i} + zp_y\mathbf{k} \times \mathbf{j} \\
&\quad + zp_z\mathbf{k} \times \mathbf{k} \\
&= xp_x \times \mathbf{0} + xp_y \times \mathbf{k} + xp_z \times (-\mathbf{j}) + yp_x \times (-\mathbf{k}) + yp_y \times \mathbf{0} + yp_z \times \mathbf{i} + zp_x \times \mathbf{j} \\
&\quad + zp_y \times (-\mathbf{i}) + zp_z \times \mathbf{0} \\
&= xp_y\mathbf{k} - xp_z\mathbf{j} - yp_x\mathbf{k} + yp_z\mathbf{i} + zp_x\mathbf{j} - zp_y\mathbf{i}
\end{aligned}$$

Collecting like terms gives

$$\mathbf{L} = (yp_z - zp_y)\mathbf{i} + (zp_x - xp_z)\mathbf{j} + (xp_y - yp_x)\mathbf{k}$$

As we saw in the previous chapter an atom has both orbital angular momentum **L** and spin angular momentum **S**. As well as adding these to give the total angular momentum

$$\mathbf{J} = \mathbf{L} + \mathbf{S}$$

they can be combined through the scalar product $\mathbf{L} \cdot \mathbf{S}$. This represents the phenomenon of spin-orbit coupling in the atom. First consider the scalar product $\mathbf{J} \cdot \mathbf{J}$

$$\begin{aligned}
\mathbf{J} \cdot \mathbf{J} &= (\mathbf{L} + \mathbf{S}) \cdot (\mathbf{L} + \mathbf{S}) \\
&= \mathbf{L} \cdot \mathbf{L} + \mathbf{L} \cdot \mathbf{S} + \mathbf{S} \cdot \mathbf{L} + \mathbf{S} \cdot \mathbf{S} \\
&= \mathbf{L} \cdot \mathbf{L} + 2\mathbf{L} \cdot \mathbf{S} + \mathbf{S} \cdot \mathbf{S}
\end{aligned}$$

since $\mathbf{L} \cdot \mathbf{S} = \mathbf{S} \cdot \mathbf{L}$.

Also

$$\begin{aligned}
\mathbf{J} \cdot \mathbf{J} &= |\mathbf{J}||\mathbf{J}| \cos 0^\circ = J^2 \\
\mathbf{L} \cdot \mathbf{L} &= |\mathbf{L}||\mathbf{L}| \cos 0^\circ = L^2 \\
\mathbf{S} \cdot \mathbf{S} &= |\mathbf{S}||\mathbf{S}| \cos 0^\circ = S^2
\end{aligned}$$

where J, L and S are used to represent the magnitudes of **J**, **L** and **S** respectively. Hence the equation above can be written as

$$J^2 = L^2 + 2\mathbf{L} \cdot \mathbf{S} + S^2$$

Subtracting $L^2 + S^2$ from both sides of this equation gives

$$2\mathbf{L} \cdot \mathbf{S} = J^2 - L^2 - S^2$$

and so

$$\mathbf{L} \cdot \mathbf{S} = \frac{1}{2}(J^2 - L^2 - S^2)$$

From this expression it is possible to use quantum mechanics to deduce the magnitude of the energy level splitting caused by spin-orbit coupling. For the sodium atom this is around 17 cm^{-1}.

Questions

1. Determine the following vectors:
   (a) $2\mathbf{a}$, where $\mathbf{a} = 3\mathbf{i} + \mathbf{j} - 2\mathbf{k}$
   (b) $3\mathbf{b}$, where $\mathbf{b} = 5\mathbf{i} - 2\mathbf{j} + 3\mathbf{k}$
   (c) $4.5\mathbf{c}$, where $\mathbf{c} = 0.7\mathbf{i} + 3.4\mathbf{j} + 2.1\mathbf{k}$

2. Determine the scalar product $\mathbf{a} \cdot \mathbf{b}$ when:
   (a) $\mathbf{a} = 3\mathbf{i} + 2\mathbf{j} + 4\mathbf{k}$ and $\mathbf{b} = -\mathbf{i} - 2\mathbf{j} + 3\mathbf{k}$
   (b) $\mathbf{a} = 3\mathbf{i} - 4\mathbf{j} - 5\mathbf{k}$ and $\mathbf{b} = -8\mathbf{i} + 6\mathbf{j} + 3\mathbf{k}$
   (c) $\mathbf{a} = 8\mathbf{j} - 7\mathbf{k}$ and $\mathbf{b} = 6\mathbf{i} + 4\mathbf{j} - 5\mathbf{k}$

3. Determine the vector product $\mathbf{a} \times \mathbf{b}$ of each pair of vectors in question 2.

4. The scalar product $\mathbf{a} \cdot \mathbf{b}$ of vectors $\mathbf{a}$ and $\mathbf{b}$ is 3.62, $|\mathbf{a}| = 2.14$ and $|\mathbf{b}| = 5.19$. Determine the angle between $\mathbf{a}$ and $\mathbf{b}$.

5. The energy E of a dipole $\boldsymbol{\mu}$ in a magnetic field $\mathbf{B}$ is given by the scalar product $E = -\boldsymbol{\mu} \cdot \mathbf{B}$
   What is the energy if a dipole of magnitude 9.274×10^{-24} J T^{-1} makes an angle of 30° with a magnetic field with $|\mathbf{B}| = 2.0$ T?

6. The angular momentum $\mathbf{L}$ of a particle of mass m is given by $\mathbf{L} = m(\mathbf{r} \times \mathbf{v})$, where the particle moves with velocity $\mathbf{v}$ and positional vector $\mathbf{r}$. If $\mathbf{r}$ is aligned such that $\mathbf{r} = a_0\mathbf{i}$, with a_0 being the Bohr radius, determine the angular momentum of an electron of mass 9.109×10^{-31} kg having $\mathbf{v} = (6 \times 10^{6}$ m s$^{-1})\mathbf{j}$ at a particular instant in time.

31. Complex Numbers

Real numbers

Most of the numbers encountered in chemistry are known as real numbers. Examples are

$$7.45 \qquad 9.2 \times 10^{-3} \qquad -1.3 \times 10^{4} \qquad 36 \qquad \frac{3}{4}$$

Note that the integers are a subset of the set of real numbers.

Imaginary numbers

The concept of imaginary numbers is rather less common. Consider the equation

$$x^2 + 1 = 0$$

Subtracting 1 from each side of this gives

$$x^2 = -1$$

Taking the square root of either side then gives

$$x = \sqrt{-1}$$

Conventionally we would say that it is not possible to take the square root of a negative number, but mathematicians choose to represent the quantity $\sqrt{-1}$ with the symbol i, which then allows the development of other branches of the subject. Notice that the symbol i is used to represent the square root of any negative number. For example, $\sqrt{-5} = \sqrt{[(-1) \times 5]} = \sqrt{-1} \times \sqrt{5} = i\sqrt{5}$.

Complex numbers

As the name suggests, a complex number is made by combining a real and an imaginary number by taking their sum. Examples of complex numbers are

$$3 + 4i = 3 + 4\sqrt{-1}$$
$$2 - 7i = 2 - 7\sqrt{-1}$$

The general complex number $(a + ib)$ is said to have real part a and imaginary part b.

Complex conjugate

The complex conjugate of $(a + ib)$ is $(a - ib)$. In general the conjugate of a complex number is obtained by replacing i by $-i$ everywhere it occurs. A common notation encountered in chemistry is that if the complex number $(a + ib)$ is denoted as z, then its conjugate is written as z^*.

The exponential of an imaginary number

The exponential of an imaginary number is the form in which most occurrences of imaginary numbers will be met in chemistry. The relationship is simply

$$e^{ikx} = \cos kx + i \sin kx$$

so that the exponential of an imaginary number ix is the complex number with real part $\cos x$ and imaginary part $\sin x$. It is also worth noting that

$$e^{-ikx} = \cos kx - i \sin kx$$

The structure factor

When X-rays are fired at a crystal, the amplitude of the scattered X-rays is defined by the structure factor $F(hkl)$, which depends on the position and nature of all the atoms in the repeat unit known as the unit cell. It can be determined from the equation

$$F(hkl) = \sum\nolimits_j f_j e^{i2\pi(hx_j+ky_j+lz_j)}$$

where f_j is known as the scattering factor of atom j having coordinates (x_j, y_j, z_j) and h, k and l are integers known as Miller indices that define each particular reflection. The symbol $\sum_j$ denotes taking the summation over all values of j. This symbol was met in Chapter 9 where it appears in the formula for calculating standard deviation.

The intensity $I(hkl)$ of each reflection is proportional to the square of the amplitude so that

$$I(hkl) \propto |F(hkl)|^2$$

Wavefunctions

A wavefunction is a mathematical state function which describes the state of a system as fully as possible, as was seen in Chapter 13. It is usually denoted by the symbol Ψ in chemistry. Since this mathematical function may be complex (i.e. contains i) it is also possible to consider its conjugate Ψ^*. This is then of use in a number of ways.

One is in calculating the probability density of a system, which gives us the probability of finding a particle at a defined position (x,y,z) and is defined as $\Psi^*\Psi$.

Complex wavefunction for particle in a box

The formula for the ground state energy for the particle in a box model was met in Chapter 13. In fact there is a series of energy levels given by the general formula

$$E = \frac{n^2h^2}{8ma^2}$$

which depend upon the quantum number n, which has values 1,2,3, ... In this formula, h is Planck's constant, m is the mass of the particle, and a is the width of the box. For each of these energy levels there is an associated wavefunction Ψ which takes the form

$$\Psi(x) = A\cos\left(\frac{n\pi x}{a}\right) + B\sin\left(\frac{n\pi x}{a}\right)$$

This clearly isn't complex, but as we saw in the previous section there may be advantages in being able to write it in such a form.

First write down the relationships

$$e^{in\pi x/a} = \cos\left(\frac{n\pi x}{a}\right) + i\sin\left(\frac{n\pi x}{a}\right)$$

$$e^{-in\pi x/a} = \cos\left(\frac{n\pi x}{a}\right) - i\sin\left(\frac{n\pi x}{a}\right)$$

Adding these expressions gives

$$e^{in\pi x/a} + e^{-in\pi x/a} = 2\cos\left(\frac{n\pi x}{a}\right)$$

where the two sine terms have added to zero. This can be rearranged to give

$$\cos\left(\frac{n\pi x}{a}\right) = \frac{e^{in\pi x/a} + e^{-in\pi x/a}}{2}$$

Subtracting the second relationship from the first gives

$$e^{in\pi x/a} - e^{in\pi x/a} = 2i\sin\left(\frac{n\pi x}{a}\right)$$

where subtraction has eliminated the cosine terms. Dividing both sides of this equation by $2i$

$$\sin\left(\frac{n\pi x}{a}\right) = \frac{e^{in\pi x/a} - e^{in\pi x/a}}{2i}$$

Now substitute for the sine and cosine terms in the original expression for $\Psi(x)$ to give

$$\Psi(x) = \frac{A}{2}\left(e^{in\pi x/a} + e^{-in\pi x/a}\right) + \frac{B}{2i}\left(e^{in\pi x/a} - e^{in\pi x/a}\right)$$

This can be rearranged to collect the like exponential terms giving

$$\Psi(x) = \left(\frac{A}{2} + \frac{B}{2i}\right)e^{in\pi x/a} + \left(\frac{A}{2} - \frac{B}{2i}\right)e^{-in\pi x/a}$$

It is now straightforward to generate the expression for $\Psi^*(x)$ by replacing i by $-i$ each time it occurs. This leads to

$$\Psi^*(x) = \left(\frac{A}{2} - \frac{B}{2i}\right)e^{-in\pi x/a} + \left(\frac{A}{2} + \frac{B}{2i}\right)e^{in\pi x/a}$$

Inspection shows that $\Psi^*(x) = \Psi(x)$, which is the expected result for a real wavefunction.

Questions

1. Give the real and imaginary parts of:
   (a) $2 + 3i$
   (b) $3 - 6i$
   (c) $4 + 7i$
   (d) $5 - 9i$
   (e) $x + iy$

2. Give the conjugate of each of the complex numbers in question 1.

3. Give the first three terms of the structure factor $F(hkl)$ in terms of separate real and imaginary parts.

4. A complex wavefunction Ψ may be written as the sum of its real R and imaginary I parts as $\Psi = R + iI$. Use this expression to show that $\Psi^*\Psi$ is real, i.e. that it has no imaginary part.

5. The wavefunctions for the $2p_x$ and $2p_y$ orbitals of the hydrogen atom can be written as

$$\Psi_{2p_x} = Ae^{-r/2a_0}r\sin\theta e^{i\phi}$$
$$\Psi_{2p_y} = Ae^{-r/2a_0}r\sin\theta e^{-i\phi}$$

   Rewrite the sum of these as a real function by using the expansions of $e^{i\phi}$and $e^{-i\phi}$ as complex numbers.

SECTION

F

Calculus

32. The Derivative

The quantity called the derivative can be thought of as the rate of change of one variable with respect to another. As an example, acceleration is the rate of change of speed with time. Graphically, the rate of change is determined by the gradient of a graph of one variable against the other.

Gradient of a straight line

The gradient of a straight line is straightforward to calculate, once the coordinates of two points A and B on the line are known. Simply determine the vertical distance Δy between the points and divide by the horizontal distance Δx between the points. In Fig. 32.1, the gradient m between points A and B is given by

$$m = \frac{\Delta y}{\Delta x} = \frac{y_2 - y_1}{x_2 - x_1}$$

So if A has coordinates (1, 2) and B has coordinates (3, 7), the gradient m will be

$$m = \frac{7-2}{3-1} = \frac{5}{2} = 2.5$$

Notice that the order of coordinates is important in this expression. If A has coordinates $(3, -1)$ while B has coordinates (1, 5), the gradient m is given by

$$m = \frac{5-(-1)}{1-3} = \frac{5+1}{-2} = \frac{6}{-2} = -3$$

i.e. the gradient is negative, so that the line now slopes in the opposite direction.

While this is an appropriate way of determining the gradient if numerical data is being dealt with, remember from Chapter 23 that the general equation of a straight line is given by $y = mx + c$, where m is the gradient and c is the intercept. Consequently, if the equation of a straight line is known its gradient m can be determined without needing to know which points it passes through.

Gradient of a curve

In order to determine the gradient of a curve at a particular point, the tangent to the curve at that point must be drawn, as shown in Fig. 32.2. Since this tangent is a straight line, its gradient can be determined from two points as outlined above. Since the tangent to the curve is different at every point on the curve, it follows that the gradient of the curve is continually changing.

In practice it is quite hard to estimate the best position in which to draw the tangent, so graphical methods do not give a very accurate value for the gradient.

In the same way that knowing the equation of a straight line allows us to determine its gradient, knowing the equation of a curve allows us to determine its gradient at any point.

The derivative

If y is a function of the single variable x, the quantity written as

$$\frac{\mathrm{d}y}{\mathrm{d}x}$$

is known as the derivative of y with respect to x. In speech we would refer to this as 'D Y by D X'. It is very important to note that both $\mathrm{d}y$ and $\mathrm{d}x$ are single entities that are never separated. You may notice a similarity between the gradient $\frac{\Delta y}{\Delta x}$ for a straight line and the gradient $\frac{\mathrm{d}y}{\mathrm{d}x}$ for a curve.

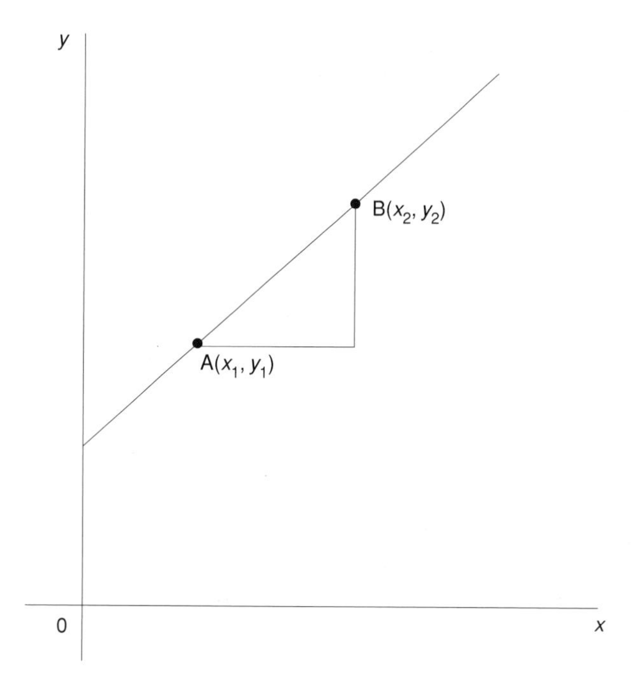

Figure 32.1 Gradient between two points A and B on a straight line

It is this derivative which gives the value of the gradient of y with respect to x, and there are rules which allow us to determine its form once we have an equation relating x and y. For example, if

$$y = x^2 - 3$$

then it will be seen in the next chapter that

$$\frac{dy}{dx} = 2x$$

It should not be surprising that another expression involving x has been obtained, since it has already been established that the gradient of a curve is constantly changing. When $x = -1$,

$$\frac{dy}{dx} = 2 \times (-1) = -2$$

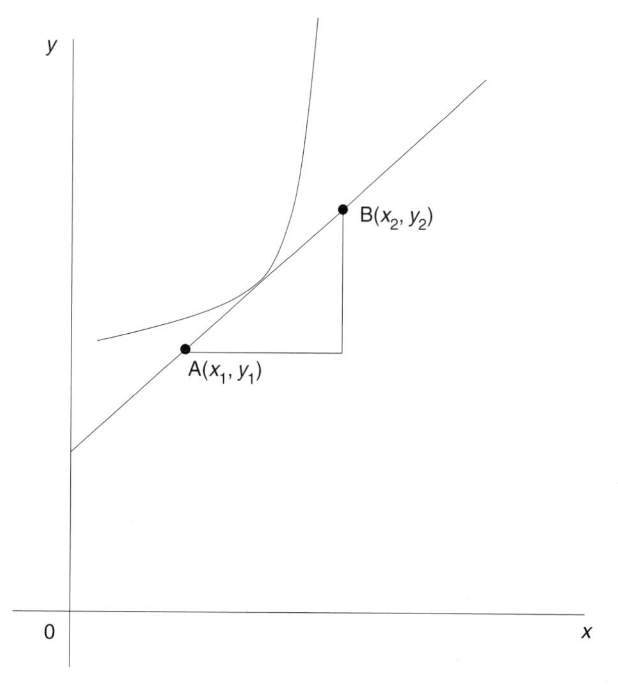

Figure 32.2 Gradient at a point on a curve

so the gradient of the curve represented by $y = x^2 - 3$ is -2 when $x = -1$.

Clapeyron equation

The Clapeyron equation gives the rate of change of pressure p with absolute temperature T during the phase changes fusion and vaporisation. If the phase change is accompanied by a change in enthalpy ΔH and a change in volume ΔV, then

$$\frac{\mathrm{d}p}{\mathrm{d}T} = \frac{\Delta H}{T \Delta V}$$

so that the gradient of a graph of p against T can be determined at any point.

When ice melts, it contracts to the extent of $1.63 \times 10^{-6}\ \mathrm{m^3\ mol^{-1}}$ and is accompanied by an enthalpy change $6.008\ \mathrm{kJ\ mol^{-1}}$. Since this takes place at an absolute temperature of 273.15 K, the derivative is given by

$$\begin{aligned}\frac{\mathrm{d}p}{\mathrm{d}T} &= \frac{6.008\ \mathrm{kJ\ mol^{-1}}}{273.15\ \mathrm{K} \times (-1.63 \times 10^{-6}\ \mathrm{m^3\ mol^{-1}})} \\ &= \frac{6.008 \times 10^3\ \mathrm{J\ mol^{-1}}}{-445.23 \times 10^{-6}\ \mathrm{K\ m^3\ mol^{-1}}} \\ &= -1.349 \times 10^7\ \mathrm{N\ m\ K^{-1}\ m^{-3}} \\ &= -1.349 \times 10^7\ \mathrm{N\ m^{-2}\ K^{-1}} \\ &= -1.349 \times 10^7\ \mathrm{Pa\ K^{-1}}\end{aligned}$$

Note that ΔV is negative in this case because of the contraction on melting.

Rates of reaction

The rate of a chemical reaction is usually described in terms of the rate of change of concentration c with respect to time t. The gradient of a graph of c against t is thus given by the derivative dc/dt.

The equation for this derivative is different for different groups of reactions. For those known as first-order reactions, it is given by the equation

$$-\frac{\mathrm{d}c}{\mathrm{d}t} = kc$$

where k is a constant known as the rate constant. The negative sign appears because as the time t increases, the concentration c decreases.

At 20°C, hydrogen peroxide decomposes in dilute sodium hydroxide according to the equation

$$2\,H_2O_{2(aq)} \rightarrow 2\,H_2O_{(l)} + O_{2(g)}$$

The reaction is first order with a rate constant k of $1.06 \times 10^{-3}\ \mathrm{min^{-1}}$.

When the concentration of hydrogen peroxide H_2O_2 is $0.010\ \mathrm{mol\ dm^{-3}}$, the derivative will be

$$\begin{aligned}\frac{\mathrm{d}c}{\mathrm{d}t} &= -1.06 \times 10^{-3}\ \mathrm{min^{-1}} \times 0.010\ \mathrm{mol\ dm^{-3}} \\ &= -1.06 \times 10^{-5}\ \mathrm{mol\ dm^{-3}\ min^{-1}}\end{aligned}$$

Questions

1. If $y = 3x^2 + 5$, it can be shown that

$$\frac{\mathrm{d}y}{\mathrm{d}x} = 6x$$

Determine the gradient of the curve $y = 3x^2 + 5$ when:
(a) $x = -4$
(b) $x = 0$
(c) $x = 2$
(d) $x = -0.5$
(e) $x = 2.42$

2. If $y = 7x^3 - 3x^2 + 9$, it can be shown that

$$\frac{dy}{dx} = 21x^2 - 6x$$

Determine the gradient of the curve $y = 7x^3 - 3x^2 + 9$ when:
(a) $x = -5$
(b) $x = 0$
(c) $x = 7$
(d) $x = -3.6$
(e) $x = 7.41$

3. If ln K for a reaction, where K is the equilibrium constant, is plotted against the absolute temperature T, then

$$\frac{d(\ln K)}{dT} = -\frac{\Delta H^{\circ}}{RT^2}$$

where ΔH° is the standard enthalpy of reaction and R is the gas constant.
Urea can be prepared using the reaction

$$2\,NH_{3(g)} + CO_{2(g)} \rightleftharpoons NH_2CONH_{2(aq)} + H_2O_{(l)}$$

for which $\Delta H^{\circ} = -119.7$ kJ mol^{-1}. What is the gradient of the graph of ln K against T at 25°C?

4. The Poiseuille equation allows viscosity η to be determined from measuring the rate of flow through a length l of a tube of radius R with a pressure difference $(p_1 - p_2)$, given by

$$\frac{dV}{dt} = \frac{(p_1 - p_2)\pi R^4}{8\eta l}$$

The viscosity of ethene at 25°C is 9.33×10^{-6} kg m^{-1} s^{-1}. What is the gradient of a graph of V against t when ethane flows through a pressure difference of 25 kPa in a tube of radius 5 mm and length 10 cm?

5. The decomposition of nitrogen dioxide

$$2\,NO_{2(g)} \rightarrow 2\,NO_{(g)} + O_{2(g)}$$

at temperatures around 350°C is a second-order reaction, obeying the equation

$$-\frac{dc}{dt} = kc^2$$

where c is the concentration of NO_2 and k is the rate constant equal to 0.775 dm^3 mol^{-1} s^{-1}. What is the gradient of the graph of c against t when $c = 0.05$ mol dm^{-3}?

33. Differentiation

The process of obtaining the derivative $\frac{dy}{dx}$ from a function $y = f(x)$ is known as differentiation. In the previous chapter it was seen that if $y = x^2 - 3$, then

$$\frac{dy}{dx} = 2x$$

An alternative way of expressing this is that if $f(x) = x^2 - 3$, then

$$\frac{df(x)}{dx} = 2x$$

In fact, the general rule for this type of function is that if $f(x) = ax^n$, then

$$\frac{df(x)}{dx} = anx^{n-1}$$

In other words, to differentiate, multiply by the power and reduce the power by 1. For example, if $f(x) = 3x^4$, then

$$\frac{df(x)}{dx} = 4 \times 3x^{4-1} = 12x^3$$

A special case of this is when $f(x) = a$, where a is a constant. Since $x^0 = 1$ (Chapter 1), a can be written as ax^0. Applying the rule above to $f(x) = ax^0$ gives

$$\frac{df(x)}{dx} = 0 \times a \times x^{0-1} = 0$$

Thus the differentiation of any constant is zero. This makes sense because graphically $y = f(x) = a$ is represented by a horizontal line, whose gradient is zero.

On the other hand, if $f(x) = ax$ is to be differentiated, the rule gives

$$\frac{df(x)}{dx} = 1 \times a \times x^{1-1} = ax^0 = a \times 1 = a$$

since $x^0 = 1$. For example, if $f(x) = 4x$, then

$$\frac{df(x)}{dx} = 1 \times 4 \times x^{1-1} = 4x^0 = 4 \times 1 = 4$$

Polynomial expressions

To differentiate a polynomial expression such as

$$f(x) = 4x^3 - 6x^2 + 12x - 4$$

simply differentiate term by term. This gives

$$\frac{df(x)}{dx} = 3 \times 4 \times x^{3-1} - 2 \times 6 \times x^{2-1} + 1 \times 12 \times x^{1-1}$$
$$= 12x^2 - 12x + 12$$

since the constant term -4 differentiates to zero.

Partial molar volumes

The partial molar volume V_1 of component 1 in a mixture is defined as the increase in total volume V when 1 mol of the component is added to a sufficiently large amount of solution that the concentration does not change. For sodium chloride, the total volume V of solution is given by

$$V = a + bm + cm^2 + dm^3$$

where m is the molality of the solution divided by the standard molality 1 mol kg^{-1}. The constants take the values $a = 1002.87$ cm^3, $b = 17.821$ cm^3 kg mol^{-1}, $c = 0.8739$ cm^3 kg^2 mol^{-2} and $d = -0.04723$ cm^3 kg^3 mol^{-3}. The quantity $\frac{dV}{dm}$ is thus the partial molar volume of sodium chloride in the solution. This is determined as

$$\frac{dV}{dm} = 1 \times b \times m^{1-1} + 2 \times c \times m^{2-1} + 3 \times d \times m^{3-1} = b + 2cm + 3dm^2$$

Thus if the molality is 0.25 mol kg^{-1},

$$\begin{aligned}\frac{dV}{dm} &= 17.821\,\text{cm}^3\,\text{kg}\,\text{mol}^{-1} + 2 \times 0.8739\,\text{cm}^3\,\text{kg}^2\,\text{mol}^{-2} \times 0.25\,\text{mol}\,\text{kg}^{-1}\\ &\quad + 3 \times (-0.047\,23)\,\text{cm}^3\,\text{kg}^3\,\text{mol}^{-3} \times (0.25\,\text{mol}\,\text{kg}^{-1})^2\\ &= 17.821\,\text{cm}^3\,\text{kg}\,\text{mol}^{-1} + 0.4370\,\text{cm}^3\,\text{kg}\,\text{mol}^{-1} - 0.0089\,\text{cm}^3\,\text{kg}\,\text{mol}^{-1}\\ &= 18.249\,\text{cm}^3\,\text{kg}\,\text{mol}^{-1}\end{aligned}$$

Heat capacity

The heat capacity C_p of methane is given by the polynomial expression

$$C_p = 23.6\,\text{J}\,\text{K}^{-1}\,\text{mol}^{-1} + 47.86 \times 10^{-3}\text{J}\,\text{K}^{-2}\,\text{mol}^{-1}T - 1.8 \times 10^5\,\text{J}\,\text{K}\,\text{mol}^{-1}T^{-2}$$

The gradient of a graph of C_p against T will be given by the derivative $\frac{dC_p}{dT}$, which is given by

$$\begin{aligned}\frac{dC_p}{dT} &= 47.86 \times 10^{-3}\,\text{J}\,\text{K}^{-2}\,\text{mol}^{-1} \times 1 \times T^{1-1} - 1.8 \times 10^5\,\text{J}\,\text{K}\,\text{mol}^{-1} \times (-2) \times T^{-2-1}\\ &= 47.86 \times 10^{-3}\,\text{J}\,\text{K}^{-2}\,\text{mol}^{-1} + 3.6 \times 10^5\,\text{J}\,\text{K}\,\text{mol}^{-1}\,T^{-3}\end{aligned}$$

At 298 K, the gradient of the graph of C_p against T is thus given by

$$\begin{aligned}\frac{dC_p}{dT} &= 47.86 \times 10^{-3}\,\text{J}\,\text{K}^{-2}\,\text{mol}^{-1} + 3.6 \times 10^5\,\text{J}\,\text{K}\,\text{mol}^{-1} \times (298\,\text{K})^{-3}\\ &= 47.86 \times 10^{-3}\,\text{J}\,\text{K}^{-2}\,\text{mol}^{-1} + \frac{3.6 \times 10^5\,\text{J}\,\text{K}\,\text{mol}^{-1}}{(298\,\text{K})^3}\\ &= 47.86 \times 10^{-3}\,\text{J}\,\text{K}^{-2}\,\text{mol}^{-1} + 13.6 \times 10^{-3}\,\text{J}\,\text{K}^{-2}\,\text{mol}^{-1}\\ &= 61.46 \times 10^{-3}\,\text{J}\,\text{K}^{-2}\,\text{mol}^{-1}\\ &= 6.146 \times 10^{-2}\,\text{J}\,\text{K}^{-2}\,\text{mol}^{-1}\end{aligned}$$

Questions

1. If a function $f(x)$ is defined by $f(x) = 3x^5 - 4x^3 + 2x$, determine an expression for $\frac{df(x)}{dx}$.

2. If a function $g(y)$ is defined by $g(y) = 2y^4 + 3y^3 - 5y^2 - y + 7$, determine an expression for $\frac{dg(y)}{dy}$.

3. Hermite polynomials form part of the solution for the quantum mechanical treatment of the harmonic oscillator. One of these polynomials is defined by

$$H_4(\xi) = 12 - 48\xi^2 + 16\xi^4$$

   Obtain an expression for $\frac{dH_4(\xi)}{d\xi}$.

4. The Laguerre polynomials form part of the solution of the R equation in the quantum mechanical treatment of the hydrogen atom. One of these polynomials is defined by

$$L_3(\rho) = 6 - 18\rho + 9\rho^2 - \rho^3$$

   Obtain an expression for $\frac{dL_3(\rho)}{d\rho}$

5. The Legendre polynomials form part of the solution of the Θ equation in the quantum mechanical treatment of the hydrogen atom. One of these polynomials is defined by

$$P_3(z) = \frac{5}{2}z^3 - \frac{3}{2}z$$

Obtain an expression for $\frac{\mathrm{d}P_3(z)}{\mathrm{d}z}$.

6. The volume V of a liquid is given by

$$V = V_0(a + bT + cT^2)$$

where V_0 is the volume of liquid at 298 K, $a = 0.85$, $b = 4.2 \times 10^{-4}\ \mathrm{K}^{-1}$, and $c = 1.67 \times 10^{-6}\ \mathrm{K}^{-2}$. Determine $\frac{\mathrm{d}V}{\mathrm{d}T}$ for this liquid at 350 K in terms of V_0.

34. Differentiation of Functions

In this chapter the derivatives are considered of four of the functions commonly encountered in chemistry. Rather than using y or $f(x)$ to denote the function each time, it will be written directly in the symbol for the derivative, as in

$$\frac{\mathrm{d}}{\mathrm{d}x}(3x^2) = 6x$$

which is a more concise expression than those used previously.

The exponential function

The general derivative of the exponential function is given by

$$\frac{\mathrm{d}}{\mathrm{d}x}(e^{ax}) = ae^{ax}$$

For example,

$$\frac{\mathrm{d}}{\mathrm{d}x}(e^{4x}) = 4e^{4x}$$

$$\frac{\mathrm{d}}{\mathrm{d}x}(e^{-2x}) = -2e^{-2x}$$

Notice that this also gives the interesting result

$$\frac{\mathrm{d}}{\mathrm{d}x}(e^{x}) = e^{x}$$

where the function and its derivative are the same.

Logarithms

The derivative of the natural logarithm is given by

$$\frac{\mathrm{d}}{\mathrm{d}x}(\ln x) = \frac{1}{x}$$

Notice that since

$$\ln ax = \ln a + \ln x$$

each term can be differentiated to give

$$\frac{\mathrm{d}}{\mathrm{d}x}(\ln ax) = \frac{\mathrm{d}}{\mathrm{d}x}(\ln a) + \frac{\mathrm{d}}{\mathrm{d}x}(\ln x) = 0 + \frac{1}{x} = \frac{1}{x}$$

since a and consequently $\ln a$ are both constants, which differentiate to zero.

Note that logarithms to the base 10 are differentiated by using the relationship

$$\log x = \frac{\ln x}{2.303}$$

i.e. by differentiating ln x and dividing the result by 2.303. This rule will be presented formally in Chapter 35.

Examples of differentiating logarithms are:

$$\frac{\mathrm{d}}{\mathrm{d}x}(\ln 5x) = \frac{1}{x}$$

$$\frac{\mathrm{d}}{\mathrm{d}x}(\log 2x) = \frac{1}{2.303}\frac{\mathrm{d}}{\mathrm{d}x}(\ln 2x) = \frac{1}{2.303} \times \frac{1}{x} = \frac{1}{2.303x}$$

Trigonometric functions

Other functions commonly differentiated in chemistry are the sine and cosine trigonometric functions. The relevant derivatives are:

$$\frac{d}{dx}[\sin(ax+b)] = a\cos(ax+b)$$

$$\frac{d}{dx}[\cos(ax+b)] = -a\sin(ax+b)$$

Note the negative sign in the second relationship.

Examples of these derivatives are:

$$\frac{d}{dx}(\sin 4x) = 4\cos 4x$$

$$\frac{d}{dx}(\cos(8x+2)) = -8\sin(8x+2)$$

$$\begin{aligned}\frac{d}{dx}(2\sin(3x-1)+4\cos 5x) &= 2\frac{d}{dx}(\sin(3x-1)) + 4\frac{d}{dx}(\cos 5x)\\ &= 2\times 3\cos(3x-1) + 4\times(-5\sin 5x)\\ &= 6\cos(3x-1) - 20\sin 5x\end{aligned}$$

The barometric distribution law

The barometric distribution law gives the distribution of atmospheric gas molecules in terms of their molar mass M, height h, absolute temperature T, the acceleration due to gravity g, and the ideal gas constant R. The pressure p at height h is given in terms of that at zero height p_0 as

$$p = p_0 e^{-Mgh/RT}$$

The rate of change of pressure with height is then given by the derivative $\frac{dp}{dh}$.

Comparing the above exponential function with e^{ax}, it can be seen that if $x = h$ then

$$a = -\frac{Mg}{RT}$$

It thus follows that

$$\frac{d}{dh}(e^{-Mgh/RT}) = -\frac{Mg}{RT}e^{-Mgh/RT}$$

and the full derivative of the expression for p is given by

$$\frac{dp}{dh} = p_0\left(-\frac{Mg}{RT}\right)e^{-Mgh/RT}$$

This isn't a particularly elegant expression, but it does contain the term $p_0 e^{-Mgh/RT}$, which the original equation defines as p. Consequently, the derivative can be rewritten as

$$\frac{dp}{dh} = -\frac{Mg}{RT}p$$

Transition state theory

Transition state theory requires the existence of an activated complex in order to overcome the activation energy of a reaction. The rate constant k_{act} for the formation of such a complex is given by

$$k_{act} = \frac{kT}{h}K^{\ddagger}$$

where k is Boltzmann's constant, T is the absolute temperature, h is Planck's constant, and $K^{\ddagger}$ is the equilibrium constant for the reaction that forms the activated complex.

The above equation can be rewritten in logarithmic form as

$$\ln k_{act} = \ln k + \ln T - \ln h + \ln K^{\ddagger}$$

To obtain the variation of $\ln k_{act}$ with temperature T, differentiate to give

$$\frac{d\ln k_{act}}{dT} = \frac{d\ln k}{dT} + \frac{d\ln T}{dT} - \frac{d\ln h}{dT} + \frac{d\ln K^{\ddagger}}{dT}$$

Since k and h are constants, $\ln k$ and $\ln h$ will also be constants which differentiate to zero. The derivative

$$\frac{d\ln T}{dT} = \frac{1}{T}$$

which gives

$$\frac{d\ln k_{act}}{dT} = \frac{1}{T} + \frac{d\ln K^{\ddagger}}{dT}$$

Questions

1. Differentiate the following expressions with respect to x:
   (a) $\ln 3x$
   (b) e^{-5x}
   (c) $\sin(4x - 7)$
   (d) $\log 7x - \cos 2x$
   (e) $e^{-x} + \sin(3x + 2) + \ln 9x$

2. The Debye–Hückel limiting law gives the mean activity coefficient $\gamma_{\pm}$ for a pair of ions with charges z_+ and z_- as

$$\log_{10} \gamma_{\pm} = -Az_+z_-\sqrt{I}$$

   where A is a constant, and I is the ionic strength/mol dm^{-3}. Obtain an expression for

$$\frac{d\log \gamma_{\pm}}{dI}$$

3. Bragg's law relates the diffraction angle θ in a crystal with planes a distance d apart to the wavelength λ and a constant n defined by:

$$n\lambda = 2d\sin\theta$$

   Obtain an expression for $\frac{d\lambda}{d\theta}$.

4. The rate constant k for a reaction is related to the absolute temperature T by the equation

$$\ln k = \ln A - \frac{E_a}{RT}$$

   where A is the pre-exponential factor, E_a is the activation energy, and R is the gas constant. Obtain an expression for $\frac{d\,lnk}{dT}$.

5. For a first-order reaction the concentration c varies with time t according to the equation $c = c_0e^{-kt}$, where c_0 is the initial concentration, and k is the rate constant. Obtain an expression for $\frac{dc}{dt}$ in terms of k and c.

35. Differentiating Combinations of Functions

Problems in chemistry sometimes require the differentiation of functions which are more complicated than those discussed so far. In the previous chapter it was seen how to differentiate a function multiplied by a constant, and sums and differences of simple functions. For completeness, these rules are formalised here, before products and quotients of functions are considered.

Differentiating a function multiplied by a constant

If the function $f(x)$ is multiplied by a constant a to form the product $af(x)$, its derivative is given by

$$\frac{\mathrm{d}}{\mathrm{dx}}(af(x)) = a\,\frac{\mathrm{d}f(x)}{\mathrm{d}x}$$

So, for example

$$\frac{\mathrm{d}}{\mathrm{dx}}(4\ln x) = 4\frac{\mathrm{d}\ln x}{\mathrm{d}x} = 4 \times \frac{1}{x} = \frac{4}{x}$$

and

$$\frac{\mathrm{d}}{\mathrm{dx}}(6\cos 2x) = 6\,\frac{\mathrm{d}}{\mathrm{dx}}(\cos 2x) = 6 \times (-2\sin 2x) = -12\sin 2x$$

Differentiating sums and differences of functions

As seen previously, sums and differences of functions are differentiated term by term. Thus to differentiate the sum $f(x) + g(x)$, we have

$$\frac{\mathrm{d}}{\mathrm{dx}}(f(x) + g(x)) = \frac{\mathrm{d}f(x)}{\mathrm{d}x} + \frac{\mathrm{d}g(x)}{\mathrm{d}x}$$

Similarly, to differentiate the difference of these functions $f(x) - g(x)$, we have

$$\frac{\mathrm{d}}{\mathrm{dx}}(f(x) - g(x)) = \frac{\mathrm{d}f(x)}{\mathrm{d}x} - \frac{\mathrm{d}g(x)}{\mathrm{d}x}$$

For example,

$$\frac{\mathrm{d}}{\mathrm{dx}}(x^3 + \sin 8x) = \frac{\mathrm{d}}{\mathrm{dx}}(x^3) + \frac{\mathrm{d}}{\mathrm{dx}}(\sin 8x) = 3x^2 + 8\cos 8x$$

and

$$\begin{aligned}\frac{\mathrm{d}}{\mathrm{dx}}(\sin 3x - \cos 4x) &= \frac{\mathrm{d}}{\mathrm{dx}}(\sin 3x) - \frac{\mathrm{d}}{\mathrm{dx}}(\cos 4x)\\ &= 3\cos 3x - (-4\sin 4x)\\ &= 3\cos 3x + 4\sin 4x\end{aligned}$$

Differentiating the product of two functions

For the product $f(x)g(x)$, the rule for differentiating is:

$$\frac{\mathrm{d}}{\mathrm{dx}}(f(x)\,g(x)) = g(x)\,\frac{\mathrm{d}f(x)}{\mathrm{d}x} + f(x)\,\frac{\mathrm{d}g(x)}{\mathrm{d}x}$$

This is best remembered in words as 'differentiate first, multiply by second, plus differentiate second, multiply by first'. For example,

$$\begin{aligned}\frac{\mathrm{d}}{\mathrm{dx}}(x^2 \ln x) &= \ln x \frac{\mathrm{d}}{\mathrm{dx}}(x^2) + x^2 \frac{\mathrm{d}}{\mathrm{dx}}(\ln x)\\ &= \ln x \times 2x + x^2 \times \frac{1}{x}\\ &= 2x \ln x + \frac{x^2}{x}\\ &= 2x \ln x + x\end{aligned}$$

A second example is

$$\begin{aligned}\frac{\mathrm{d}}{\mathrm{dx}}(\sin 2x \cos 2x) &= \cos 2x \frac{\mathrm{d}}{\mathrm{dx}}(\sin 2x) + \sin 2x \frac{\mathrm{d}}{\mathrm{dx}}(\cos 2x)\\ &= \cos 2x \times 2\cos 2x + \sin 2x \times (-2\sin 2x)\\ &= 2\cos^2 2x - 2\sin^2 2x\end{aligned}$$

Differentiating the quotient of two functions

If $f(x)$ is divided by $g(x)$, the rule for differentiating the quotient is:

$$\frac{\mathrm{d}}{\mathrm{dx}}\left(\frac{f(x)}{g(x)}\right) = \frac{g(x)\,\frac{\mathrm{d}f(x)}{\mathrm{d}x} - f(x)\,\frac{\mathrm{d}g(x)}{\mathrm{d}x}}{[g(x)]^2}$$

In words this can be remembered as 'differentiate top, multiply by bottom, minus differentiate bottom multiplied by top, divide by bottom squared'. For example,

$$\begin{aligned}\frac{\mathrm{d}}{\mathrm{dx}}\left(\frac{x^2}{\ln x}\right) &= \frac{\ln x \frac{\mathrm{d}}{\mathrm{dx}}(x^2) - x^2 \frac{\mathrm{d}}{\mathrm{dx}}(\ln x)}{(\ln x)^2}\\ &= \frac{\ln x \times 2x - x^2 \times \frac{1}{x}}{(\ln x)^2}\\ &= \frac{2x \ln x - \frac{x^2}{x}}{(\ln x)^2}\\ &= \frac{2x \ln x - x}{(\ln x)^2}\end{aligned}$$

There are various ways in which this answer could be expressed, none of them particularly elegant. Fortunately there are few occasions in chemistry when differentiating a quotient is required.

Variation of enthalpy with temperature

The enthalpy change ΔH for a reaction is a quantity that chemists use routinely. However, it is based on a precise definition of enthalpy H, which can be measured at the beginning and end of a reaction. This definition is:

$$H = U + pV$$

where U is the internal energy of the system, p is the pressure, and V is the volume. The temperature variation of the enthalpy is then given by

$$\frac{\mathrm{d}H}{\mathrm{d}T} = \frac{\mathrm{d}}{\mathrm{d}T}(U + pV) = \frac{\mathrm{d}U}{\mathrm{d}T} + \frac{\mathrm{d}}{\mathrm{d}T}(pV)$$

In order to differentiate pV with respect to T, the rule for differentiating a product can be used, which gives

$$\frac{\mathrm{d}}{\mathrm{d}T}(pV) = V\frac{\mathrm{d}p}{\mathrm{d}T} + p\frac{\mathrm{d}V}{\mathrm{d}T}$$

so that

$$\frac{\mathrm{d}H}{\mathrm{d}T} = \frac{\mathrm{d}U}{\mathrm{d}T} + V\frac{\mathrm{d}p}{\mathrm{d}T} + p\frac{\mathrm{d}V}{\mathrm{d}T}$$

Techniques such as this are frequently used to generate relationships in thermodynamics.

Gibbs–Helmholtz equation

The Gibbs–Helmholtz equation equation gives us the variation of the change in Gibbs free energy, ΔG, with temperature T. An important part of its derivation requires the differentiation of the quantity $\Delta G/T$. It is important to realise that ΔG does depend upon T, so that this is an example of differentiating a quotient. If ΔG did not vary with temperature, then the task would be simpler i.e. differentiating a constant divided by a variable. It may also seem strange to be differentiating a quantity involving the Δ symbol, but in this case ΔG is simply a function of T.

To differentiate $\Delta G/T$, apply the rule for differentiating a quotient to give

$$\begin{aligned}\frac{\mathrm{d}}{\mathrm{d}T}\left(\frac{\Delta G}{T}\right) &= \frac{T\,\frac{\mathrm{d}\Delta G}{\mathrm{d}T} - \Delta G\,\frac{\mathrm{d}}{\mathrm{d}T}(T)}{T^2} \\ &= \frac{T\,\frac{\mathrm{d}\Delta G}{\mathrm{d}T} - \Delta G \times 1}{T^2} \\ &= \frac{T\,\frac{\mathrm{d}\Delta G}{\mathrm{d}T}}{T^2} - \frac{\Delta G}{T^2} \\ &= \frac{T}{T^2}\frac{\mathrm{d}\Delta G}{\mathrm{d}T} - \frac{\Delta G}{T^2} \\ &= \frac{1}{T}\frac{\mathrm{d}\Delta G}{\mathrm{d}T} - \frac{\Delta G}{T^2}\end{aligned}$$

There are a few other steps involved in the derivation of the final Gibbs–Helmholtz equation, but this is probably the trickiest.

Questions

1. Differentiate the following functions with respect to x:
   (a) $4x^5$
   (b) $x^3 - x^2 + x - 9$
   (c) $3\ln x - 4\sin 2x$
   (d) $6x - e^{-3x} + \ln 5x$
   (e) $5/x^3 - 2x + \ln 8x$

2. Differentiate the following functions with respect to x:
   (a) $x \ln x$
   (b) $x^2 \mathrm{e}^{-2x}$
   (c) $3x \sin 2x$
   (d) $4xe^x + x$
   (e) $7x^2 \cos 4x + xe^x$

3. Differentiate the following functions with respect to x:
   (a) $x/\ln x$
   (b) $\sin x/x^2$
   (c) $e^x/\ln 2x$
   (d) $\sin x/\cos 3x$
   (e) $x \ln x/\sin x$

4. Ostwald's dilution law gives the relationship between the equilibrium constant K for a dissociation reaction and the molar conductivity Λ of the resulting solution of concentration c, which is

$$K = \frac{c\left(\frac{\Lambda}{\Lambda_0}\right)}{1 - \left(\frac{\Lambda}{\Lambda_0}\right)}$$

   where Λ_0 is the molar conductivity at infinite dilution. Differentiate K with respect to the quantity Λ/Λ_0.

5. In the statistical mechanical treatment of the Boltzmann distribution law the derivative

$$\frac{\mathrm{d}}{\mathrm{d}n_i}(n_i \ln n_i - n_i)$$

   occurs. Determine an expression for this derivative.

36. Higher-Order Differentiation

In the previous chapters it has been shown that differentiating a function gives us a second function in the same variables. It follows that differentiating this second function should be possible, and the process repeated a number of times in certain cases.

Consider the polynomial function

$$f(x) = 2x^3 - 4x^2 + 3x + 1$$

Differentiating this gives

$$\frac{df(x)}{dx} = 6x^2 - 8x + 3$$

using the standard rule for differentiating polynomial expressions. Differentiating this again gives

$$12x - 8$$

which is called the second derivative of $f(x)$, written as $\frac{d^2f(x)}{dx^2}$ and called 'D 2 $f(x)$ by D X squared'.

The relationship between $\frac{df(x)}{dx}$ and $\frac{d^2f(x)}{dx^2}$ can be expressed as

$$\frac{d^2f(x)}{dx^2} = \frac{d}{dx}\left(\frac{df(x)}{dx}\right)$$

Differentiating the expression a third time gives $\frac{d^3f(x)}{dx^3}$, and so on.

Suppose that

$$f(x) = \ln 3x$$

This differentiates to give

$$\frac{df(x)}{dx} = \frac{d}{dx}(\ln 3x) = \frac{1}{x}$$

which can also be written as x^{-1}. Consequently

$$\frac{d^2f(x)}{dx^2} = \frac{d}{dx}(x^{-1}) = -1 \times x^{-2} = -\frac{1}{x^2}$$

If $f(x) = 3e^{2x}$, the derivative is

$$\frac{df(x)}{dx} = 6e^{2x} \quad \text{and so} \quad \frac{d^2f(x)}{dx^2} = 12e^{2x}$$

Particle in a one-dimensional box

The wavefunction Ψ for the particle within a one-dimensional box of length a can be found by solving the Schrödinger equation:

$$-\frac{h^2}{8\pi^2 m}\frac{d^2\Psi}{dx^2} = E\Psi$$

where m is the mass of the particle, E is its energy, and h is Planck's constant. The particle is constrained to move only in the x direction.

The second-order derivative, $\frac{d^2\Psi}{dx^2}$, can be isolated by multiplying both sides of the equation by $-8\pi^2m/h^2$ to give

$$\frac{d^2\Psi}{dx^2} = -\frac{8\pi^2 mE}{h^2}\Psi$$

The wavefunction for such a system is usually obtained by making an intelligent guess of the form of function required for Ψ and substituting into this equation to obtain the value of any unknown constants, along with using any boundary condition information.

A suitable wavefunction is of the form

$$\Psi = \left(\frac{2}{a}\right)^{1/2} \sin\left(\frac{n\pi x}{a}\right)$$

where n can take the values $1, 2, 3, \ldots$. It is important to realise that n, π and a are constants so that

$$\frac{d\Psi}{dx} = \left(\frac{2}{a}\right)^{1/2} \left(\frac{n\pi}{a}\right) \cos\left(\frac{n\pi x}{a}\right)$$

and

$$\frac{d^2\Psi}{dx^2} = \left(\frac{2}{a}\right)^{1/2} \times \left[-\left(\frac{n\pi}{a}\right)^2 \sin\left(\frac{n\pi x}{a}\right)\right] = -\left(\frac{2}{a}\right)^{1/2} \left(\frac{n\pi}{a}\right)^2 \sin\left(\frac{n\pi x}{a}\right)$$

Substituting this into the Schrödinger equation above now gives

$$-\left(\frac{2}{a}\right)^{1/2} \left(\frac{n\pi}{a}\right)^2 \sin\left(\frac{n\pi x}{a}\right) = -\frac{8\pi^2 mE}{h^2}\left(\frac{2}{a}\right)^{1/2} \sin\left(\frac{n\pi x}{a}\right)$$

The terms belonging to the definition of Ψ

$$\Psi = \left(\frac{2}{a}\right)^{1/2} \sin\left(\frac{n\pi x}{a}\right)$$

cancel on either side of the expression to give

$$\left(\frac{n\pi}{a}\right)^2 = \frac{8\pi^2 mE}{h^2} \quad \text{or} \quad \frac{n^2\pi^2}{a^2} = \frac{8\pi^2 mE}{h^2}$$

Cancelling π^2 on each side gives

$$\frac{n^2}{a^2} = \frac{8mE}{h^2}$$

and finally multiplying each side by $h^2/8m$ gives the energy levels of the particle as

$$E = \frac{n^2h^2}{8ma^2}$$

Quantum mechanics of the hydrogen atom

Chapter 28 referred to the need to solve an equation in the variable θ for the hydrogen atom. Part of this solution includes Legendre polynomials of degree l, which are defined as

$$P_l(z) = \frac{1}{2^l l!}\frac{d^l}{dz^l}(z^2 - 1)^l$$

Considering $P_2(z)$, means $l = 2$, so that

$$l! = 2 \times 1 = 2 \qquad 2^l = 2^2 = 4 \qquad \frac{d^l}{dz^l} = \frac{d^2}{dz^2}$$

and

$$(z^2 - 1)^2 = (z^2 - 1)(z^2 - 1) = z^4 - 2z^2 + 1$$

This gives

$$\frac{d}{dz}(z^4 - 2z^2 + 1) = 4z^3 - 4z \qquad \text{and} \qquad \frac{d^2}{dz^2}(z^4 - 2z^2 + 1) = 12z^2 - 4$$

Combining these together now gives

$$\begin{aligned} P_2(z) &= \frac{1}{4 \times 2}(12z^2 - 4) \\ &= \frac{1}{8}(12z^2 - 4) \\ &= \frac{12}{8}z^2 - \frac{4}{8} \\ &= \frac{3}{2}z^2 - \frac{1}{2} \end{aligned}$$

Questions

1. Determine $d^2 f(x)/dx^2$ when:
   (a) $f(x) = 4x^3 - 3x^2 + x - 5$
   (b) $f(x) = 6x^4 - 3x^2$
   (c) $f(x) = 9x^2 + 3x - 1$

2. Determine $d^2 g(y)/dy^2$ when:
   (a) $g(y) = \ln 4y$
   (b) $g(y) = 2e^{-4y}$
   (c) $g(y) = \ln 3y + e^{2y}$

3. Determine $d^2 h(z)/dz^2$ when:
   (a) $h(z) = \sin 3z$
   (b) $h(z) = \cos(4z + 1)$
   (c) $h(z) = \sin 2z + \cos 2z$

4. The wavefunction for the 1s orbital of the hydrogen atom is given by

$$\Psi_{1s} = \left(\frac{1}{\pi}\right)^{1/2}\left(\frac{1}{a_0}\right)^{3/2} e^{-r/a_0}$$

   where a_0 is the Bohr radius, and r is the distance of the electron from the nucleus. Obtain an expression for $\frac{d^2\Psi_{1s}}{dr^2}$.

5. The volume V of a solution can be expressed in terms of the amount n of salt present in 1kg of water by the equation

$$V = a + bn + cn^{3/2} + en^2$$

   Determine the value of $\frac{d^2V}{dn^2}$, when $n = 0.25$ mol for sodium chloride for which $a = 1002.96\ \text{cm}^3$, $b = 16.6253\ \text{cm}^3\ \text{mol}^{-1}$, $c = 1.7738\ \text{cm}^3\ \text{mol}^{-3/2}$ and $e = 0.1194\ \text{cm}^3\ \text{mol}^{-2}$.

37. Stationary Points

So far it has been shown that the derivative gives the gradient of the tangent to a specified curve at any point. At the turning or stationary points, also known as maxima and minima, the gradient of the tangent will be zero. Consequently, if the points at which the derivative is zero can be determined, the turning points of the curve will be known.

For example, if

$$f(x) = 3x^2 - 9x + 2$$

the derivative is

$$\frac{\mathrm{d}f(x)}{\mathrm{d}x} = 6x - 9$$

Setting

$$\frac{\mathrm{d}f(x)}{\mathrm{d}x} = 0$$

gives

$$6x - 9 = 0$$

Adding 9 to both sides gives the equation

$$6x = 9$$

and dividing by 6

$$x = \frac{9}{6} = \frac{3}{2}$$

Determining the nature of the stationary point

Determining the nature of the stationary point is done by looking at the value of $\frac{\mathrm{d}^2 f(x)}{\mathrm{d}x^2}$; if it is negative the stationary point is a maximum whereas if it is positive the stationary point is a minimum. If the second derivative is zero the stationary point may be a maximum, a minumum, or a point of inflexion as shown in Fig. 37.1. Its nature in this case can only be determined by considering the sign of the gradient each side of the stationary point.

In the example above, the second derivative is

$$\frac{\mathrm{d}^2 f(x)}{\mathrm{d}x^2} = \frac{\mathrm{d}}{\mathrm{d}x}(6x - 9) = 6$$

Since this is positive, the stationary point at $x = 3/2$ is a miniumum.

Consider now the function

$$f(x) = 2x^3 - 3x + 3$$

Differentiating gives

$$\frac{\mathrm{d}f(x)}{\mathrm{d}x} = 6x^2 - 3$$

Setting this to zero gives

$$6x^2 - 3 = 0$$

Removing the common factor 3 allows this to be rewritten using brackets as

$$3(2x^2 - 1) = 0$$

Consequently

$$2x^2 = 1 \qquad \text{so that} \qquad x^2 = \frac{1}{2}$$

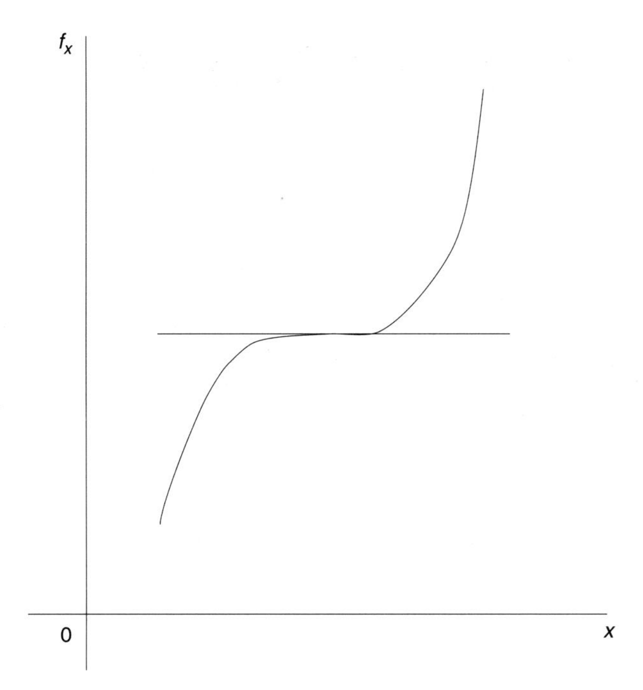

Figure 37.1 A point of inflection

It is important to realise that this has two solutions,

$$x = -\frac{1}{\surd 2} \quad \text{and} \quad x = \frac{1}{\surd 2}$$

Calculating the second derivative gives

$$\frac{d^2 f(x)}{dx^2} = 12x$$

Consequently, when $x = -\frac{1}{\surd 2}$,

$$\frac{d^2 f(x)}{dx^2} = 12 \times (-\frac{1}{\surd 2}) = -\frac{12}{\surd 2}$$

so this is a maximum.

When $x = \frac{1}{\surd 2}$,

$$\frac{d^2 f(x)}{dx^2} = 12 \times \frac{1}{\surd 2} = \frac{12}{\surd 2}$$

so this is a minimum.

Note that it is not always necessary to obtain the final numerical value of $\frac{d^2 f(x)}{dx^2}$ in order to determine whether it is positive or negative.

A final example is that of

$$f(x) = x^3 - 4$$

The derivative is

$$\frac{df(x)}{dx} = 3x^2$$

This can only be zero when $x = 0$, which thus denotes the stationary point.

The second derivative is

$$\frac{d^2 f(x)}{dx^2} = 6x$$

This will clearly be zero when $x = 0$, so in this case we need to test further.

When $x = -1$,

$$\frac{df(x)}{dx} = 3 \times (-1)^2 = 3 \times 1 = 3$$

When $x = 1$,

$$\frac{df(x)}{dx} = 3 \times 1^2 = 3 \times 1 = 3$$

Since the gradient is positive both before and after the stationary point, this must be a point of inflection. The same would be true if the gradient was negative both before and after the stationary point.

The Lennard–Jones potential

This has been widely used to study the interactions between uncharged molecules, and gives the potential energy of interaction V in terms of the separation r:

$$V(r) = -\frac{A}{r^6} + \frac{B}{r^{12}}$$

In order to differentiate this, it is helpful to rewrite the expression as

$$V(r) = -Ar^{-6} + Br^{-12}$$

The derivative can now be determined as

$$\begin{aligned}\frac{\mathrm{d}V(r)}{\mathrm{d}r} &= -A(-6r^{-7}) + B(-12r^{-13})\\ &= 6Ar^{-7} - 12Br^{-13}\\ &= \frac{6A}{r^7} - \frac{12B}{r^{13}}\end{aligned}$$

In order to determine the stationary points it is helpful to rewrite it with the common factor $1/r^7$ outside the brackets to give

$$\frac{\mathrm{d}V(r)}{\mathrm{d}r} = \frac{1}{r^7}\left(6A - \frac{12B}{r^6}\right)$$

For this expression to be zero

$$6A - \frac{12B}{r^6} = 0$$

which rearranges to

$$6A = \frac{12B}{r^6}$$

Multiplying both sides by r^6 gives

$$6Ar^6 = 12B$$

and dividing by $6A$ leads to

$$r^6 = \frac{12B}{6A} = \frac{2B}{A}$$

Returning to the expression for $\frac{\mathrm{d}V(r)}{\mathrm{d}r}$ and differentiating again gives

$$\begin{aligned}\frac{\mathrm{d}^2V(r)}{\mathrm{d}r^2} &= \frac{\mathrm{d}}{\mathrm{d}r}(6Ar^{-7} - 12Br^{-13})\\ &= 6A(-7r^{-8}) - 12B(-13r^{-14})\\ &= -42Ar^{-8} + 156Br^{-14}\\ &= -\frac{42A}{r^8} + \frac{156B}{r^{14}}\\ &= \frac{1}{r^8}\left(\frac{156B}{r^6} - 42A\right)\end{aligned}$$

Substituting directly for r^6 gives

$$\begin{aligned}\frac{\mathrm{d}^2V(r)}{\mathrm{d}r^2} &= \frac{1}{r^8}\left(156B \times \frac{A}{2B} - 42A\right)\\ &= \frac{1}{r^8}(78A - 42A)\\ &= \frac{36A}{r^8}\end{aligned}$$

Since r and A are both positive $\frac{\mathrm{d}^2V(r)}{\mathrm{d}r^2}$ will also be positive, so this stationary point is a minimum.

The helium atom

The variation theorem can be used to provide a quantum mechanical determination of the energy $\bar{E}$ of the helium atom. This is expressed in terms of the effective nuclear charge Z' experienced by each of the electrons as

$$\bar{E} = [-2(Z')^2 + \frac{27}{4}Z']E_H$$

where E_H is the Hartree (Appendix 4). In order to find the value of Z' that minimises the energy, we need to set the derivative $\frac{d\bar{E}}{dZ'}$ equal to zero. This derivative is easily determined since the expression above is actually a simple polynomial.

$$\frac{d\bar{E}}{dZ'} = \left[-4Z' + \frac{27}{4}\right]E_H = 0$$

Since E_H is not zero it follows that the quantity in the bracket must be. Setting this equal to zero and rearranging gives

$$4Z' = \frac{27}{4}$$

and dividing both sides of the equation by 4

$$Z' = \frac{27}{16}$$

Returning to the expression for $\frac{d\bar{E}}{dZ'}$ and differentiating a second time gives

$$\frac{d^2\bar{E}}{dZ'^2} = \frac{d}{dZ}\left(-4Z' + \frac{27}{4}\right)E_H = -4E_H$$

Since E_H is negative, $\frac{d^2\bar{E}}{dZ'^2}$ is positive, and so $Z' = 27/16$ does correspond to the minimum value of $\bar{E}$.

Questions

1. For what value of x does the function $f(x) = 3x^2 - 6x + 7$ have a stationary point?
2. For what values of y does the function $g(y) = \ln 2y - 2y^2$ have a stationary point?
3. Identify the positions and nature of the stationary points of the function $f(x) = 4x^3 - 6x^2 + 1$.
4. Identify the positions of the stationary points of the function $f(x) = \sin(3x - 5)$ in the range $0 \leq x \leq 2\pi$.
5. The potential energy V between two ions a distance r apart can be described by the equation

$$V(r) = -\frac{A}{r} + \frac{B}{r^6}$$

Obtain an expression for the position of the stationary point of this function. Show that this is a minimum.

6. The energy E for the 1 s orbital in the helium atom is given by the equation

$$E = \frac{e^2}{a_0}\left(Z^2 - \frac{27}{8}Z\right)$$

where e is the electronic charge, a_0 the Bohr radius, and Z the effective nuclear charge seen by the electron. Calculate the minimum energy in terms of e and a_0.

38. Partial Differentiation

We have seen so far how to differentiate a function of a single variable. However, many functions in chemistry have two or more variables, and in these cases the rate of change with respect to each one still needs to be measured.

This is done by effectively fixing the value of each term. Consider, for example, the function

$$f(x, y) = 2x + 3y - 4xy$$

In order to find out how $f(x, y)$ varies with respect to x, y could be given a fixed value, say 4, and the resulting expression differentiated with respect to x

$$f(x, 4) = 2x + 3 \times 4 - 4x \times 4$$

or

$$f(x, 4) = 2x + 12 - 16x = 12 - 14x$$

This is effectively what is done when the partial derivative of $f(x, y)$ with respect to x is determined; the process is called partial differentiation. The partial derivative is denoted by

$$\left(\frac{\partial f(x, y)}{\partial x}\right)_y$$

The subscript outside the bracket denotes the variable which is kept constant. We do not give y a numerical value, but simply remember that it remains constant. It is sometimes helpful to rewrite the original expression with all the constant terms in brackets, so

$$f(x, y) = (2)x + (3y) - (4y)x$$

Then applying the standard rules of differentiation gives

$$\left(\frac{\partial f(x, y)}{\partial x}\right)_y = 2 - 4y$$

The derivative of $(3y)$ is zero in this case because y is being held constant. It is also possible to determine the partial derivative of $f(x, y)$ with respect to y. Rewriting the original expression as

$$f(x, y) = (2x) + (3)y - (4x)y$$

then

$$\left(\frac{\partial f(x, y)}{\partial y}\right)_x = 3 - 4x$$

This time the derivative of $(2x)$ is zero because x is being held constant.

A more challenging function would be $f(x, y) = 3x^2y + \ln(xy)$. With y held constant, this could be written as

$$f(x, y) = (3y)x^2 + \ln((y)x)$$

so that

$$\left(\frac{\partial f(x, y)}{\partial x}\right)_y = (3y)\,2x + \frac{1}{x} = 6xy + \frac{1}{x}$$

If x is held constant,

$$f(x, y) = (3x^2)y + \ln((x)y)$$

and so

$$\left(\frac{\partial f(x, y)}{\partial y}\right)_x = 3x^2 + \frac{1}{y}$$

The differential

Once we have expressions for the partial derivatives with respect to each variable in the function, it is possible to combine them to give the overall variation in the function. This is known as the differential, which would be written as d*f* for a function $f(x, y)$. The general expression for the differential is

$$df = \left(\frac{\partial f}{\partial x}\right)_y dx + \left(\frac{\partial f}{\partial y}\right)_x dy$$

which can be read as the change *df* in f produced by changes *dx* in x and *dy* in y. Note that $f(x, y)$ has been abbreviated to f for reasons of clarity. The subscripts x and y are also often omitted where the meaning of the symbols for the partial derivative are obvious.

In the example above the differential would be

$$df = \left(6xy + \frac{1}{x}\right) dx + \left(3x^2 + \frac{1}{y}\right) dy$$

When $x = 1$ and $y = 2$, this would become

$$df = \left(6 \times 1 \times 2 + \frac{1}{1}\right) dx + \left(3 \times 1^2 + \frac{1}{2}\right) dy = 13dx + \frac{7}{2}dy$$

This then allows us to predict the behaviour of $f(x, y)$ when there are small changes around $x = 1$ and $y = 2$.

Higher order partial derivatives

Once we have obtained, say, $\left(\frac{\partial f}{dx}\right)_y$ it is possible to differentiate it again as in the case of a full derivative. This second-order partial derivative is denoted as $\left(\frac{\partial^2 f}{\partial x^2}\right)_y$ although it is often written in abbreviated form as $\frac{\partial^2 f}{\partial x^2}$ if it is clear which variable or variables are kept constant. In the above example where

$$f(x, y) = 3x^2y + \ln(xy)$$

and

$$\left(\frac{\partial f(x, y)}{\partial x}\right)_y = 6xy + \frac{1}{x} \qquad \left(\frac{\partial f(x, y)}{\partial y}\right)_x = 3x^2 + \frac{1}{y}$$

we have

$$\begin{aligned}\left(\frac{\partial^2 f(x, y)}{\partial x^2}\right)_y &= \left[\frac{\partial}{\partial x}\left(6xy + \frac{1}{x}\right)\right]_y \\ &= \left[\frac{\partial}{\partial x}(6xy + x^{-1})\right]_y \\ &= 6y - x^{-2} \\ &= 6y - \frac{1}{x^2}\end{aligned}$$

and

$$\begin{aligned}\left(\frac{\partial^2 f(x, y)}{\partial y^2}\right)_x &= \left[\frac{\partial}{\partial y}\left(3x^2 + \frac{1}{y}\right)\right]_x \\ &= \left[\frac{\partial}{\partial y}(3x^2 + y^{-1})\right]_x \\ &= -y^{-2} \\ &= -\frac{1}{y^2}\end{aligned}$$

However, it is also possible to determine two other second-order partial derivatives. The first of these would be denoted by $\frac{\partial^2 f(x,y)}{\partial y \partial x}$, where the partial derivative $\frac{\partial f}{\partial x}$ is further differentiated, but this time with respect to the variable y. In the example above, this gives

$$\frac{\partial^2 f(x,y)}{\partial y\,\partial x} = \left[\frac{\partial}{\partial y}\left(6xy + \frac{1}{x}\right)\right] = 6x$$

The final partial derivative is $\frac{\partial^2 f(x,y)}{\partial x\,\partial y}$, where the partial derivative $\frac{\partial f}{\partial y}$ is now differentiated with respect to x. In our example, this gives

$$\frac{\partial^2 f(x,y)}{\partial x\,\partial y} = \left[\frac{\partial}{\partial x}\left(3x^2 + \frac{1}{y}\right)\right]_y$$

$$= \left[\frac{\partial}{\partial x}\left(3x^2 + y^{-1}\right)\right]_y = 6x$$

Notice that $\frac{\partial^2 f}{\partial x\,\partial y} = \frac{\partial^2 f}{\partial y\,\partial x}$. This relationship will be true for many of the functions we meet.

Euler's chain rule

If z is a function of x and y, the following relationship applies

$$\left(\frac{\partial x}{\partial y}\right)_z \left(\frac{\partial y}{\partial z}\right)_x \left(\frac{\partial z}{\partial x}\right)_y = -1$$

As a very simple example of this consider

$$z(x,y) = 3xy$$

This can be rearranged to give

$$x = \frac{z}{3y} \quad \text{and} \quad y = \frac{z}{3x}$$

The necessary partial derivatives are thus

$$\left(\frac{\partial x}{\partial y}\right)_z = \frac{z}{3}(-y^{-2}) = -\frac{z}{3y^2}$$

$$\left(\frac{\partial y}{\partial z}\right)_x = \frac{1}{3x}$$

$$\left(\frac{\partial z}{\partial x}\right)_y = 3y$$

Furthermore, since $z = 3xy$, we can rewrite

$$\left(\frac{\partial x}{\partial y}\right)_z = -\frac{3xy}{3y^2} = -\frac{x}{y}$$

Substituting into Euler's chain rule now gives

$$-\frac{x}{y} \times \frac{1}{3x} \times 3y = -\frac{3xy}{3xy} = -1$$

as required.

The van der Waals equation

The behaviour of a gas can be described in terms of its pressure p, volume V, and absolute temperature T, in terms of the van der Waals equation

$$\left(p + \frac{an^2}{V^2}\right)(V - nb) = nRT$$

where n is the amount of gas, and a and b are constants.

In order to obtain an expression for the differential $\mathrm{d}p$ of the pressure, both sides of this equation can be divided by $(V - nb)$ to give

$$\left(p + \frac{an^2}{V^2}\right) = \frac{nRT}{V - nb}$$

Subtracting $\frac{an^2}{V^2}$ from both sides then gives

$$p = \frac{nRT}{V - nb} - \frac{an^2}{V^2}$$

The partial derivative with respect to T is straightforward to calculate since it only appears in the first term where it is multiplied by a series of constants. Thus

$$\left(\frac{\partial p}{\partial T}\right)_V = \frac{nR}{V - nb}$$

The partial derivative with respect to V is somewhat trickier. For the first term we use the rule for differentiating a quotient (Chapter 35) while for the second we write $1/V^2$ as V^{-2}. The partial derivative is thus

$$\begin{aligned}\left(\frac{\partial p}{\partial V}\right)_T &= \frac{0 \times (V - nb) - 1 \times nRT}{(V - nb)^2} - an^2(-2V^{-3}) \\ &= -\frac{nRT}{(V - nb)^2} + \frac{2an^2}{V^3}\end{aligned}$$

The overall differential dp can now be written as

$$\mathrm{d}p = \left(\frac{nR}{V - nb}\right)\mathrm{d}T + \left(\frac{2an^2}{V^3} - \frac{nRT}{(V - nb)^2}\right)\mathrm{d}V$$

Travelling wave in a stretched string

This model is often used to introduce the behaviour of waves in a classical system, in advance of meeting quantum mechanical systems. The general differential equation of wave motion in one dimension is

$$\frac{\partial^2 y}{\partial x^2} = \frac{1}{u^2}\left(\frac{\partial^2 y}{\partial t^2}\right)$$

where y is the transverse displacement of the string, x is the distance along the string, u is a constant, and t is time. A possible trial function for y is

$$y = A \sin(ax + bt + c)$$

where A, a, b and c are constants. Calculating the first- and second-order partial derivatives with respect to both x and t gives

$$\begin{aligned}\frac{\partial y}{\partial x} &= Aa \cos(ax + bt + c) \\ \frac{\partial^2 y}{\partial x^2} &= -Aa^2 \sin(ax + bt + c) \\ \frac{\partial y}{\partial t} &= Ab \cos(ax + bt + c) \\ \frac{\partial^2 y}{\partial t^2} &= -Ab^2 \sin(ax + bt + c)\end{aligned}$$

Substituting the two second-order partial derivatives into the equation for wave motion gives

$$-Aa^2 \sin(ax + bt + c) = \frac{1}{u^2}\left(-Ab^2 \sin(ax + bt + c)\right)$$

The term $-A \sin(ax + bt + c)$ can be cancelled to leave

$$a^2 = \frac{1}{u^2} \times b^2 = \frac{b^2}{u^2}$$

so that

$$u^2 = \frac{b^2}{a^2} \quad \text{and} \quad u = \pm\frac{b}{a}$$

Questions

1. The function $f(x, y)$ is defined by $f(x, y) = 5x^2y - 3xy - 4xy^2$. Obtain expressions for $\frac{\partial f}{\partial x}$ and $\frac{\partial f}{\partial y}$.

2. The function $g(x, y)$ is defined by

$$g(x, y) = x^2 + \ln\left(\frac{x}{y}\right)$$

   Obtain expressions for $\frac{\partial g}{\partial x}$ and $\frac{\partial g}{\partial y}$.

3. The function $h(r, s)$ is defined by $h(r, s) = e^{-r} + \sin(r + 2s)$. Obtain expressions for $\frac{\partial h}{\partial r}$ and $\frac{\partial h}{\partial s}$.

4. The ideal gas equation can be written in the form

$$p = \frac{nRT}{V}$$

   where p is the pressure, V is the volume, and T is the absolute temperature. The amount of gas is n and R is the ideal gas constant. Obtain expressions for $\left(\frac{\partial p}{\partial T}\right)_V$, $\left(\frac{\partial p}{\partial V}\right)_T$ and the differential $\mathrm{d}p$.

5. The entropy of mixing ΔS of two gases can be found from the equation

$$\Delta S = -R(x_1 \ln x_1 + x_2 \ln x_2)$$

   where x_1 and x_2 are the respective mole fractions, and R is the ideal gas constant. Obtain expressions for $\frac{\partial \Delta S}{\partial x_1}$, $\frac{\partial \Delta S}{\partial x_2}$ and the differential.

39. Integration

Integration is essentially the reverse process of differentiation. If we differentiate the function

$$f(x) = x^2 - 3$$

we obtain the derivative $2x$. Reversing this statement leads us to conclude that if we integrate the function

$$f(x) = 2x$$

we obtain $x^2 - 3$. This is called the integral of the function $2x$.

In fact, things are not quite as straightforward as this. Since the derivative of any constant is zero, functions such as

$$f(x) = x^2 - 7$$
$$f(x) = x^2 + 2$$
$$f(x) = x^2 + 10$$

could be differentiated and they would all have $2x$ as their derivative. In more general terms, if the symbol C is used to denote a constant then the derivative of the function

$$f(x) = x^2 + C$$

is $2x$ and the integral of $2x$ is $x^2 + C$. Representing the process using mathematical symbols gives

$$\int 2x \, \mathrm{d}x = x^2 + C$$

which gives 'the integral of $2x$ with respect to x'. The constant C is known as the constant of integration.

Definite integration

The fact that integrals produce an unknown constant C is not usually a problem. In chemistry we are usually concerned with finding what is known as the definite integral. One example of this based on the above result would be the determination of

$$\int_1^2 2x \, \mathrm{d}x$$

which would be called 'the definite integral of $2x$ with respect to x from $x = 1$ to $x = 2$'. The numbers on the integral sign are known as integration limits; the lower limit at the bottom and the upper limit at the top. To determine this we evaluate the value of the integral at the upper limit. This will be

$$x^2 + C = 2^2 + C = 4 + C$$

We then evaluate the integral at the lower limit as

$$x^2 + C = 1^2 + C = 1 + C$$

Then subtract the lower limit from the upper limit to give

$$(4 + C) - (1 + C) = 3$$

Note that the constant of integration C cancels and a numerical answer is obtained. This will always be the case.

This working would be set out more concisely as follows:

$$\int_1^2 2x \, dx = \left[x^2\right]_1^2$$
$$= \left[2^2\right] - \left[1^2\right] = 4 - 1 = 3$$

Note the use of the square brackets and the fact that the constant of integration is conventionally omitted when dealing with limits.

Second-order kinetics

When analysing the kinetics of a second-order reaction, the integrals

$$\int_{c_0}^{c} \frac{1}{c^2} \, dc \quad \text{and} \quad \int_0^t dt$$

need to be determined. Notice that in both cases symbols are used as limits; this is a frequently used technique in order to obtain useful general expressions in which numerical values are substituted at a later stage. The limit c can be thought of as a particular value of the variable c in the expression being integrated; similarly limit t is a particular value of the variable t. The lower limit c_0 is the initial value of c when $t = 0$. Since

$$\int \frac{1}{x^2} \, dx = -\frac{1}{x} + C$$

the definite integral required is

$$\int_{c_0}^{c} \frac{1}{c^2} \, dc = \left[-\frac{1}{c}\right]_{c_0}^{c}$$
$$= \left[-\frac{1}{c}\right] - \left[-\frac{1}{c_0}\right]$$
$$= \frac{1}{c_0} - \frac{1}{c}$$

Note the need to treat the negative signs carefully in this expression.

In order to determine $\int_0^t dt$ we need to be aware that

$$\int dx = x + C$$

Consequently

$$\int_0^t dt = [t]_0^t = t - 0 = t$$

Once determined as above, the individual integrals can be combined to give the overall equation for second-order kinetics. This was met earlier, in part (b) of question 4 in Chapter 23.

Clausius–Clapeyron equation

The Clausius–Clapeyron equation gives the variation of the vapour pressure p of a liquid with absolute temperature T. To derive the relationship involves integration of the expression

$$\int_{p_1}^{p_2} \frac{1}{p} \, dp$$

Since

$$\int \frac{1}{x}\mathrm{d}x = \ln x + C$$

this expression becomes

$$\int_{p_1}^{p_2} \frac{1}{p}\mathrm{d}p = [\ln p]_{p_1}^{p_2}$$
$$= \ln p_2 - \ln p_1$$
$$= \ln\left(\frac{p_2}{p_1}\right)$$

since $(\ln a - \ln b) = \ln(a/b)$, as shown in Chapter 19.

Integration of this type of function is common in chemistry.

Questions

1. Evaluate the integral $\int_1^4 x^3\,\mathrm{d}x$ if $\int x^3\,\mathrm{d}x = \frac{x^4}{4} + C$.

2. Evaluate the integral $\int_{-2}^{2} \frac{1}{x^2}\,\mathrm{d}x$ if $\int \frac{1}{x^2}\,\mathrm{d}x = -\frac{1}{x} + C$.

3. Evaluate the integral $\int_0^{2\pi} \sin 2x\,\mathrm{d}x$ if $\int \sin 2x\,\mathrm{d}x = -\frac{1}{2}\cos 2x + C$.

4. An equation to relate concentration c to time t for a reaction obeying zero-order kinetics can be obtained by evaluating

$$\int_{c_0}^{c} \mathrm{d}c = -k\int_0^t \mathrm{d}t$$

where c_0 is the initial concentration. Integrate this expression using the fact that $\int \mathrm{d}x = x + C$.

5. An expression for the barometric pressure p at height z can be obtained from

$$\int_{p_0}^{p} \frac{\mathrm{d}p}{p} = -\frac{mg}{RT}\int_0^z \mathrm{d}z$$

where p_0 is the pressure at the base, m is the mass of fluid, g is the acceleration due to gravity, R is the gas constant, and T is the absolute temperature. Integrate this expression using the relationships

$$\int \frac{1}{x}\,\mathrm{d}x = \ln x + C \quad \text{and} \quad \int \mathrm{d}x = x + C$$

40. Integration of Functions

This chapter considers the mechanics of how the common functions are integrated, giving the rules for each and some examples.

Polynomials

$$\int x^n \mathrm{d}x = \frac{x^{n+1}}{n+1} + C$$

In words, to integrate an expression of the form x^n, increase the power by one and divide by the new power. For example,

$$\int x^3 \,\mathrm{d}x = \frac{x^{3+1}}{3+1} + C = \frac{x^4}{4} + C$$

$$\int x^{-2} \,\mathrm{d}x = \frac{x^{-2+1}}{-2+1} + C = \frac{x^{-1}}{-1} + C = -\frac{1}{x} + C$$

Note that this rule doesn't work if $n = -1$ since we would then be dividing by zero. This case is dealt with below.

If the expression to be integrated is multiplied by a constant, the resulting integral is multiplied by the same constant. For example,

$$\begin{aligned}\int 3x^5 \,\mathrm{d}x &= 3\int x^5 \,\mathrm{d}x \\ &= 3\left(\frac{x^{5+1}}{5+1} + C\right) \\ &= 3\left(\frac{x^6}{6} + C\right) \\ &= \frac{3x^6}{6} + 3C \\ &= \frac{x^6}{2} + C'\end{aligned}$$

where $C' = 3C$. Note that in practice the constant C would simply be added at the end after working through other aspects of the integration.

A constant on its own can also be integrated in this way. Since any number raised to the power zero is 1, then

$$a = a \times 1 = ax^0$$

Applying the rule then gives

$$\int a \,\mathrm{d}x = \int ax^0 \,\mathrm{d}x = a\int x^0 \,\mathrm{d}x = a\left(\frac{x^{0+1}}{1+1}\right) + C = ax^1 + C = ax + C$$

For example,

$$\int \mathrm{d}x = x + C$$

$$\int 4 \,\mathrm{d}x = 4\int \mathrm{d}x = 4x + C$$

With a sum or difference of terms, simply integrate term by term. For example,

$$\int (5x^3 - 3x + 4) \,\mathrm{d}x = \frac{5x^4}{4} - \frac{3x^2}{2} + 4x + C$$

The reciprocal

As we saw above, an exception to the previous rule is the reciprocal, i.e. when $n = -1$. In this case

$$\int x^{-1}\,\mathrm{d}x = \int \frac{1}{x}\,\mathrm{d}x = \ln x + C$$

As an example

$$\int \frac{1}{2x}\,\mathrm{d}x = \frac{1}{2}\int \frac{1}{x}\,\mathrm{d}x = \frac{1}{2}\ln x + C$$

The exponential function

The rule for integrating an exponential function is

$$\int e^{ax}\,\mathrm{d}x = \frac{e^{ax}}{a} + C$$

For example,

$$\int 3e^{-4x}\,\mathrm{d}x = 3\int e^{-4x}\,\mathrm{d}x = 3\left(\frac{e^{-4x}}{-4}\right) + C = -\frac{3}{4}e^{-4x} + C$$

Using the rules of indices this could also be written as

$$-\frac{3}{4e^{4x}} + C$$

Trigonometric functions

The rules for integrating the sine and cosine functions are

$$\int \cos ax\,\mathrm{d}x = \frac{1}{a}\sin ax + C$$

$$\int \sin ax\,\mathrm{d}x = -\frac{1}{a}\cos ax + C$$

Note the negative sign which appears in the second of these. Examples are

$$\int 3\cos 2x\,\mathrm{d}x = 3\int \cos 2x\,\mathrm{d}x = 3 \times \frac{1}{2}\sin 2x + C = \frac{3}{2}\sin 2x + C$$

$$\int 2\sin 5x\,\mathrm{d}x = 2\int \sin 5x\,\mathrm{d}x = 2 \times \left(-\frac{1}{5}\right)\cos 5x + C = -\frac{2}{5}\cos 5x + C$$

First order kinetics

The following equation needs to be solved in order to give the integrated rate equation for a first-order reaction:

$$-\frac{\mathrm{d}c}{\mathrm{d}t} = kc$$

where t is the time, c is the concentration, and k is the rate constant. This equation can be rearranged to group the two variables c and t on either side by multiplying by $-dt$ and dividing by c to give

$$\frac{\mathrm{d}c}{c} = -k\,\mathrm{d}t$$

This can be integrated once the appropriate limits have been chosen. If we wish to find the concentration c after time t, these should be our upper limits. If c_0 is the initial concentration then this is the lower limit when $t = 0$. Integrating both sides of this equation within these limits now gives

$$\int_{c_0}^{c} \frac{\mathrm{d}c}{c} = -k\int_{0}^{t} \mathrm{d}t$$

Since $\int \frac{1}{x}\,\mathrm{d}x = \ln x + C$ and $\int \mathrm{d}x = x$,

$$[\ln c]_{c_0}^{c} = -k\,[t]_0^t$$

as in question 4 in Chapter 39. Putting in these limits gives

$$\ln c - \ln c_0 = -k(t-0)$$

so that

$$\ln c - \ln c_0 = -kt$$

Alternatively, using the rules for combining logarithms from Chapter 19, this could be written as

$$\ln \frac{c_0}{c} = kt$$

Normalising a wavefunction

A possible wavefunction Ψ for the first excited state in the particle in a box model, which we have met on several previous occasions, is

$$\Psi = x(2x-a)(x-a)$$

where a is the length of the box in the x-direction. The probability of finding the particle is given by integrating the product $\Psi^*\Psi$; in this case since Ψ is real (see Chapter 31), Ψ^2 needs to be integrated. If the limits of integration coincide with the boundaries of the box, then the probability of finding the particle must be 1. Consequently

$$\int_0^a \Psi^*\Psi\,\mathrm{d}x = 1$$

For the wavefunction given in this case, the brackets can be multiplied to give

$$\Psi = 2x^3 - 3ax^2 + a^2x$$

This can then be squared with a little work. Multiplying $(2x^3 - 3\mathrm{ax}^2 + a^2x)$ by $(2x^3 - 3ax^2 + a^2x)$ and collecting like terms gives

$$\Psi^2 = 4x^6 - 12ax^5 + 13a^2x^4 - 6a^3x^3 + a^4x^2$$

Consequently

$$\begin{aligned}\int_0^a \Psi^*\Psi\,\mathrm{d}x &= \int_0^a \left(4x^6 - 12ax^5 + 13a^2x^4 - 6a^3x^3 + a^4x^2\right)\mathrm{d}x \\ &= \left[\frac{4x^7}{7} - \frac{12ax^6}{6} + \frac{13a^2x^5}{5} - \frac{6a^3x^4}{4} + \frac{a^4x^3}{3}\right]_0^a \\ &= \left[\frac{4a^7}{7} - \frac{12aa^6}{6} + \frac{13a^2a^5}{5} - \frac{6a^3a^4}{4} + \frac{a^4a^3}{3}\right] - [0] \\ &= \left[\frac{4a^7}{7} - \frac{12a^7}{6} + \frac{13a^7}{5} - \frac{6a^7}{4} + \frac{a^7}{3}\right]\end{aligned}$$

A little experimentation shows that all of these terms can be expressed in terms of fractions of 210, so that

$$\begin{aligned}\int \Psi^*\Psi\,\mathrm{d}x &= \frac{(120-420+546-315+70)\,a^7}{210} \\ &= \frac{a^7}{210}\end{aligned}$$

In order to normalise the wavefunction, it needs to be multiplied by $\sqrt{\left(\frac{210}{a^7}\right)}$, in order to ensure that $\int_0^a \Psi^2\mathrm{d}x = 1$. This is the normalisation constant.

Questions

1. Evaluate the following integrals:
   (a) $\int x^9 \, dx$
   (b) $\int 3x^{-6} \, dx$
   (c) $\int \frac{2}{3x} \, dx$
   (d) $\int 2e^{-5x} \, dx$
   (e) $\int \cos 3x \, dx$

2. Evaluate the following:
   (a) $\int_0^6 (5x^3 - 2x^2 + x + 6) \, dx$
   (b) $\int_{-1}^3 (x^2 - 2x + 1) \, dx$
   (c) $\int_{-4}^0 (3x^4 - 4x^2 - 7) \, dx$

3. Evaluate the following:
   (a) $\int_1^5 \frac{1}{2x} \, dx$
   (b) $\int_0^1 e^{3x} \, dx$
   (c) $\int_0^\pi \sin 4x \, dx$

4. The enthalpy change ΔH associated with a change of temperature from T_1 to T_2 can be found from the equation

$$\Delta H = \int_{T_1}^{T_2} C_p \, dT$$

   The heat capacity C_p can be expressed in the form

$$C_p = a + bT + cT^{-2}$$

   For nitrogen gas, $a = 28.58$ J K^{-1} mol^{-1}, $b = 3.76 \times 10^{-3}$ J K^{-2} mol^{-1}, and $c = -5.0 \times 10^4$ J K mol^{-1}. Calculate ΔH when the temperature rises from 298 K to 318 K.

5. The variation of equilibrium constant K with temperature T can be found from the equation

$$\int_{K_1}^{K_2} \frac{dK}{K} = \frac{\Delta H^{\ominus}}{R} \int_{T_1}^{T_2} \frac{dT}{T^2}$$

   Integrate this expression to give K_1 and K_2 in terms of T_1 and T_2. $\Delta H^{\ominus}$ is the standard enthalpy change and R is the gas constant.

41. Integration Techniques

There are specific rules for differentiating specified combinations of functions (see Chapter 35), but this is not the case for integration, which is generally a less straightforward process. In this chapter three techniques will be considered that are used for integrating more complicated functions in chemistry; in general it isn't necessary to decide which method to use as this is well established.

Integration by parts

The rule for differentiating a product of two functions was met in Chapter 35. If the functions are u and v then

$$\frac{\mathrm{d}}{\mathrm{d}x}(uv) = v\,\frac{\mathrm{d}u}{\mathrm{d}x} + u\,\frac{\mathrm{d}v}{\mathrm{d}x}$$

This can be rearranged by subtracting $v\,\frac{\mathrm{d}u}{\mathrm{d}x}$ from both sides to give

$$u\,\frac{\mathrm{d}v}{\mathrm{d}x} = \frac{\mathrm{d}}{\mathrm{d}x}(uv) - v\,\frac{\mathrm{d}u}{\mathrm{d}x}$$

Integrating this expression with respect to x term by term now gives

$$\int u\,\frac{\mathrm{d}v}{\mathrm{d}x}\,\mathrm{d}x = \int \frac{\mathrm{d}}{\mathrm{d}x}(uv)\,\mathrm{d}x - \int v\,\frac{\mathrm{d}u}{\mathrm{d}x}\,\mathrm{d}x$$

Since integrating the derivative of a function must give back the original function,

$$\int u\,\frac{\mathrm{d}v}{\mathrm{d}x}\,\mathrm{d}x = uv - \int v\,\frac{\mathrm{d}u}{\mathrm{d}x}\,\mathrm{d}x$$

which is the equation for integration by parts.

Suppose we wish to determine the integral

$$\int x\cos x\,\mathrm{d}x$$

We need to compare this with $\int u\,\frac{\mathrm{d}v}{\mathrm{d}x}\,\mathrm{d}x$, bearing in mind that we will need to differentiate u and integrate $\frac{\mathrm{d}v}{\mathrm{d}x}$. In this case we therefore set $u = x$ and $\frac{\mathrm{d}v}{\mathrm{d}x} = \cos x$, as integrating the latter does not give us a more complicated function to deal with. If $u = x$ it follows that $\frac{\mathrm{d}u}{\mathrm{d}x} = 1$, and if $\frac{\mathrm{d}v}{\mathrm{d}x} = \cos x$, it follows that $v = \sin x$. Note that the constant of integration can be included at the end of the problem if required.

Substituting values into the general equation above gives

$$\begin{aligned}\int x\cos x\,\mathrm{d}x &= x\sin x - \int \sin x \times 1\,\mathrm{d}x\\ &= x\sin x - \int \sin x\,\mathrm{d}x\\ &= x\sin x + \cos x + C\end{aligned}$$

This result can be verified by differentiation. In the case of a definite integral we would now substitute the values of x for the upper and lower limits and take the difference.

Integration by substitution

Suppose we wish to calculate $\int_1^2 (2x+3)^5\mathrm{d}x$. This can be done by making an appropriate substitution, in this case $u = 2x + 3$. Differentiating this gives $\mathrm{d}u/\mathrm{d}x = 2$, so that $\mathrm{d}x = \mathrm{d}u/2$. Note also that when $x = 1$, $u = 2 \times 1 + 3 = 5$, and when $x = 2$, $u = 2 \times 2 + 3 = 7$.

Now substituting into the original expression gives

$$\int_5^7 u^5 \frac{\mathrm{d}u}{2} = \frac{1}{2}\int_5^7 u^5 \mathrm{d}u$$

$$= \frac{1}{2}\left[\frac{u^6}{6}\right]_5^7 = \frac{1}{12}\left[7^6 - 5^6\right]$$

$$= \frac{1}{12}[117\,649 - 15\,625] = \frac{102\,024}{12}$$

$$= 8502$$

Note that the expression to be integrated must be fully rewritten in terms of the new variables including the final dx, and that the limits of the integration must also correspond to the new variable.

Partial fractions

Although not actually an integration technique, certain expressions can be rewritten in terms of partial fractions to make integration possible. Suppose we have the expression

$$\frac{3}{(x+1)(x-2)}$$

This can be written as

$$\frac{A}{x+1} + \frac{B}{x-2}$$

where A and B are constants to be determined. Comparing the two expressions gives

$$\frac{3}{(x+1)(x-2)} = \frac{A(x-2) + B(x+1)}{(x+1)(x-2)}$$

where the lower part of each fraction is the same, and the terms on the top and the bottom cancel in the second version to give the previous result. Equating the top of these fractions gives

$$3 = A(x-2) + B(x+1)$$

Setting $x = 2$, then

$$3 = A(2-2) + B(2+1) \quad \text{or} \quad 3 = 3B$$

so that $B = 1$.

Setting $x = -1$ gives

$$3 = A(-1-2) + B(-1+1) \quad \text{or} \quad 3 = -3A$$

and $A = -1$.

Consequently

$$\frac{3}{(x+1)(x-2)} = \frac{-1}{(x+1)} + \frac{1}{(x-2)}$$

This is an easier expression to integrate than the original one.

Calculating an expectation value

As seen previously, the wavefunction Ψ for a particle constrained within a box of length a is given by

$$\Psi = \left(\frac{2}{a}\right)^{1/2} \sin\left(\frac{n\pi x}{a}\right)$$

It can be shown that the expectation value $<x>$ for the position of the particle is given by

$$<x> = \frac{2}{a}\int_0^a \frac{x}{2}\left[1 - \cos\left(\frac{2n\pi x}{a}\right)\right]\mathrm{d}x$$

This expands to give

$$< x > = \frac{1}{a}\int_0^a x\,\mathrm{d}x - \frac{1}{a}\int_0^a x\cos\left(\frac{2n\pi x}{a}\right)\mathrm{d}x$$

The first integral is evaluated easily as

$$\frac{1}{a}\int_0^a x\,\mathrm{d}x = \frac{1}{a}\left[\frac{x^2}{2}\right]_0^a$$

$$= \frac{1}{2a}\left[a^2 - 0^2\right] = \frac{a^2}{2a} = \frac{a}{2}$$

The second can be evaluated using integration by parts. In the equation

$$\int u\frac{\mathrm{d}v}{\mathrm{d}x}\,\mathrm{d}x = uv - \int v\frac{\mathrm{d}u}{\mathrm{d}x}\,\mathrm{d}x$$

set $u = x$ and $\mathrm{d}v/\mathrm{d}x = \cos\ (2n\pi x/a)$ so that $\mathrm{d}u/\mathrm{d}x = 1$ and

$$v = \frac{a}{2n\pi}\sin\left(\frac{2n\pi x}{a}\right)$$

Then

$$\int_0^a x\cos\left(\frac{2n\pi x}{a}\right)\mathrm{d}x = \left[x\frac{a}{2n\pi}\sin\left(\frac{2n\pi x}{a}\right)\right]_0^a - \frac{a}{2n\pi}\int_0^a \sin\left(\frac{2n\pi x}{a}\right)\times 1\,\mathrm{d}x$$

$$= \left[\frac{xa}{2n\pi}\sin\left(\frac{2n\pi x}{a}\right) - \frac{a}{2n\pi}\left(-\frac{a}{2n\pi}\right)\cos\left(\frac{2n\pi x}{a}\right)\right]_0^a$$

$$= \left[\frac{xa}{2n\pi}\sin\left(\frac{2n\pi x}{a}\right) + \frac{a^2}{4n^2\pi^2}\cos\left(\frac{2n\pi x}{a}\right)\right]_0^a$$

$$= \left[\frac{a^2}{2n\pi}\sin\left(\frac{2n\pi a}{a}\right) + \frac{a^2}{4n^2\pi^2}\cos\left(\frac{2n\pi a}{a}\right)\right] - \left[\frac{a^2}{4n^2\pi^2}\cos 0\right]$$

$$= 0 + \frac{a^2}{4n^2\pi^2}\times 1 - \frac{a^2}{4n^2\pi^2} = 0$$

since $\sin 2n\pi = 0$, $\cos 2n\pi = 1$ and $\cos 0 = 1$. Consequently only the first term of the integral contributes, and this was evaluated earlier as $a/2$. Consequently $<x> = a/2$.

Second-order kinetics

When the reactants react in other than a 1:1 ratio the analysis is less straightforward. Consider the reaction

$$aA + bB \rightarrow \text{products}$$

where a and b are stoichiometric coefficients. The rate for this reaction can be defined as

$$rate = -\frac{1}{a}\frac{\mathrm{d[A]}}{\mathrm{d}t} = -\frac{1}{b}\frac{\mathrm{d[B]}}{\mathrm{d}t} = k[\mathrm{A}][\mathrm{B}]$$

Writing the concentrations [A] and [B] in terms of the initial concentrations $[\mathrm{A}]_0$ and $[\mathrm{B}]_0$ and the amount of reactant x consumed, gives

$$[\mathrm{A}] = [\mathrm{A}]_0 - ax \quad \text{and} \quad [\mathrm{B}] = [\mathrm{B}]_0 - bx$$

so that

$$\frac{\mathrm{d[A]}}{\mathrm{d}t} = -a\frac{\mathrm{d}x}{\mathrm{d}t} \quad \text{and} \quad \frac{\mathrm{d[B]}}{\mathrm{d}t} = -b\frac{\mathrm{d}x}{\mathrm{d}t}$$

and

$$rate = -\frac{1}{a}\left(-a\frac{\mathrm{d}x}{\mathrm{d}t}\right) = \frac{\mathrm{d}x}{\mathrm{d}t}$$

$$rate = -\frac{1}{b}\left(-b\frac{\mathrm{d}x}{\mathrm{d}t}\right) = \frac{\mathrm{d}x}{\mathrm{d}t}$$

The rate equation can now be written as

$$\frac{dx}{dt} = k([A]_0 - ax)([B]_0 - bx)$$

Dividing each side by the brackets and multiplying by dt gives

$$\int_0^x \frac{dx}{([A]_0 - ax)([B]_0 - bx)} = k \int_0^t dt$$

The left-hand side can be integrated if we rewrite it in terms of partial fractions. Setting

$$\begin{aligned}\frac{1}{([A]_0 - ax)([B]_0 - bx)} &= \frac{X}{([A]_0 - ax)} + \frac{Y}{([B]_0 - bx)} \\ &= \frac{X([B]_0 - bx) + Y([A]_0 - ax)}{([A]_0 - ax)([B]_0 - bx)}\end{aligned}$$

Equating the first and third fraction gives

$$1 = X([B]_0 - bx) + Y([A]_0 - ax)$$

Setting $x = [B]_0/b$, then $1 = Y([A]_0 - a\,[B]_0/b)$ so that

$$\begin{aligned}Y &= \frac{1}{[A]_0 - \frac{a[B]_0}{b}} \\ &= \frac{1}{\frac{b[A]_0 - a[B]_0}{b}} \\ &= \frac{b}{b[A]_0 - a[B]_0}\end{aligned}$$

Setting $x = [A]_0/a$, then $1 = X([B]_0 - bx)$ so that

$$\begin{aligned}X &= \frac{1}{[B]_0 - \frac{b[A]_0}{a}} \\ &= \frac{1}{\frac{a[B]_0 - b[A]_0}{a}} \\ &= \frac{a}{a[B]_0 - b[A]_0}\end{aligned}$$

The equation to be integrated now becomes

$$\frac{a}{(a[B]_0 - b[A]_0)} \int_0^x \frac{dx}{([A]_0 - ax)} + \frac{b}{(b[A]_0 - a[B]_0)} \int_0^x \frac{dx}{([B]_0 - bx)} = k \int_0^t dt$$

The necessary integrations on the left of this equation can be performed by using the substitution method. If we set $u = [A]_0 - ax$, $du/dx = -a$ so $dx = -du/a$ and

$$\begin{aligned}\int \frac{dx}{[A]_0 - x} &= \int \frac{1}{u}\left(-\frac{du}{a}\right) \\ &= -\frac{1}{a} \int \frac{du}{u} \\ &= -\frac{1}{a} \ln u \\ &= -\frac{1}{a} \ln([A]_0 - ax)\end{aligned}$$

Similarly

$$\int \frac{dx}{[B]_0 - bx} = -\frac{1}{b} \ln([B]_0 - bx)$$

Returning to our expression obtained using partial fractions and substituting the values of the above integrals gives

$$\frac{a}{(a[\mathrm{B}]_0 - b[\mathrm{A}]_0)}\left[-\frac{1}{a}\ln([\mathrm{A}]_0 - ax)\right]_0^x + \frac{b}{(b[\mathrm{A}]_0 - a[\mathrm{B}]_0)}\left[-\frac{1}{b}\ln([\mathrm{B}]_0 - bx)\right]_0^x = k\,[t]_0^t$$

This can be simplified by cancelling terms top and bottom and inserting the limits to give

$$\frac{1}{(a[\mathrm{B}]_0 - b[\mathrm{A}]_0)}[-\{\ln([\mathrm{A}]_0 - ax) - \ln[\mathrm{A}]_0\}] + \frac{1}{(b[\mathrm{A}]_0 - a[\mathrm{B}]_0)}[-\{\ln([\mathrm{B}]_0 - bx) - \ln[\mathrm{B}]_0\}]$$
$$= kt$$

Multiplying the second fraction top and bottom by -1 on the left-hand side and factorising gives

$$\frac{1}{a[\mathrm{B}]_0 - b[\mathrm{A}]_0}\left\{\ln\frac{[\mathrm{B}]_0 - bx}{[\mathrm{B}]_0} - \ln\frac{[\mathrm{A}]_0 - ax}{[\mathrm{A}]_0}\right\} = kt$$

The variable x can now be eliminated from the expression by looking back at the original definitions so that

$$\frac{1}{a[\mathrm{B}]_0 - b[\mathrm{A}]_0}\left\{\ln\frac{[\mathrm{B}]}{[\mathrm{B}]_0} - \ln\frac{[\mathrm{A}]}{[\mathrm{A}]_0}\right\} = kt$$

Using the rules for combining logarithms from Chapter 19, we have

$$\frac{1}{a[\mathrm{B}]_0 - b[\mathrm{A}]_0}\ln\frac{[\mathrm{A}]_0[\mathrm{B}]}{[\mathrm{A}][\mathrm{B}]_0} = kt$$

Questions

1. Use integration by parts to evaluate:
   (a) $\int x \sin x \, dx$
   (b) $\int xe^x \, dx$
   (c) $\int x \ln x \, dx$
   (d) $\int_1^3 xe^{3x} \, dx$
   (e) $\int_0^{\pi/2} 3x \cos 2x \, dx$

2. Evaluate the following by using an appropriate substitution:
   (a) $\int_{-1}^{4} (x-2)^6 \, dx$
   (b) $\int_{-\pi}^{\pi/2} \sin(4x+1) \, dx$
   (c) $\int_0^2 3xe^{x^2} \, dx$

3. Evaluate the following using partial fractions and appropriate substitutions:
   (a) $\int_5^7 \frac{3x}{(x+1)(x-4)} \, dx$
   (b) $\int \frac{2}{x(x+1)} \, dx$
   (c) $\int_5^{10} \frac{2x-3}{(x+5)(x-2)} \, dx$

4. The total probability of finding an electron in the hydrogen atom is related to the integral

$$\int r^2 e^{-2r/a_0} \, dr$$

where r is the distance of the electron from the nucleus and a_0 is the Bohr radius. Determine this integral using integration by parts.

5. The rotational partition function q_r of a molecule is given by the equation

$$q_r = \int_0^\infty (2J + 1)\exp\left(\frac{-J(J+1)h^2}{8\pi^2 IkT}\right) \mathrm{d}J$$

where J is the rotational quantum number, h is Planck's constant, I is the moment of inertia, k is Boltzmann's constant, and T is the absolute temperature. Use the substitution $u = J(J+1)$ to obtain an expression for this integral.

Appendices

Appendix 1

SI Prefixes

Multiplier	Name of SI prefix	Symbol
10^{-18}	atto	a
10^{-15}	femto	f
10^{-12}	pico	p
10^{-9}	nano	n
10^{-6}	micro	μ
10^{-3}	milli	m
10^{-2}	centi	c
10^{-1}	deci	d
10^{3}	kilo	k
10^{6}	mega	M
10^{9}	giga	G
10^{12}	tera	T

Appendix 2

SI Unit Conversions

SI unit	Conversion	Base unit
N		$kg\ m\ s^{-2}$
Pa	$N\ m^{-2}$	$kg\ m^{-1}\ s^{-2}$
J	$N\ m$	$kg\ m^{2}\ s^{-2}$
V	$J\ C^{-1}$	$kg\ m^{2}\ s^{-3}\ A^{-1}$
Hz		s^{-1}
T	$V\ S\ m^{-2}$	$kg\ s^{-2}\ A^{-1}$

Appendix 3

Non-SI Units

Non-SI unit	SI equivalent
atm	101 325 Pa
bar	10^{5} Pa
eV	1.60×10^{-19} J
Å	10^{-10} m
cal	4.184 J

Appendix 4

Values of Physical Constants

Physical constant	Symbol	Value
Avogadro constant	L	6.022×10^{23} mol^{-1}
electronic charge	e	1.602×10^{-19} C
gas constant	R	8.314 J K^{-1} mol^{-1}
Boltzmann's constant	k	1.381×10^{-23} J K^{-1}
Planck's constant	h	6.626×10^{-34} J s
speed of light	c	2.998×10^{8} m s^{-1}
Faraday	F	9.649×10^{4} C mol^{-1}
permittivity of free space	ε_0	8.854×10^{-12} F m^{-1}
electron mass	m_e	9.109×10^{-31} kg
Rydberg constant	R	1.097×10^{7} m^{-1}
Hartree	E_H	-4.360×10^{-18} J
Bohr radius	a_0	5.292×10^{-11} m
acceleration due to gravity	g	9.807 m s^{-2}
pi	π	3.142

Appendix 5

Table of t -Values

	Confidence level			
Degrees of freedom	**90%**	**95%**	**97.5%**	**99%**
2	2.92	4.30	6.21	9.93
3	2.35	3.18	4.18	5.84
4	2.13	2.78	3.50	4.60
5	2.02	2.57	3.16	4.03
6	1.94	2.45	2.97	3.71
7	1.89	2.36	2.84	3.50
8	1.86	2.31	2.75	3.36
9	1.83	2.26	2.69	3.25
10	1.81	2.23	2.63	3.17
11	1.80	2.20	2.59	3.11
12	1.78	2.18	2.56	3.05
13	1.77	2.16	2.53	3.01
14	1.76	2.14	2.51	2.98
15	1.75	2.13	2.49	2.95
16	1.75	2.12	2.47	2.92
17	1.74	2.11	2.46	2.90
18	1.73	2.10	2.45	2.88
19	1.73	2.09	2.43	2.86
20	1.72	2.09	2.42	2.85
21	1.72	2.08	2.41	2.83
22	1.72	2.07	2.41	2.82
23	1.71	2.07	2.40	2.81
24	1.71	2.06	2.39	2.80
25	1.71	2.06	2.38	2.79

Solutions to Problems

Chapter 1

1. a. $a^3 \times a^5 = a^{3+5} = a^8$
   b. $x^2 \times x^6 = x^{2+6} = x^8$
   c. $y^4 \times y^3 = y^{4+3} = y^7$
   d. $b^6/b^3 = b^{6-3} = b^3$
   e. $x^{10}/x^2 = x^{10-2} = x^8$

2. a. $(c^4)^3 = c^{4\times 3} = c^{12}$
   b. $z^0 = 1$
   c. $1/y^4 = y^{-4}$
   d. $x^5 \times x^{-5} = x^{5-5} = x^0 = 1$
   e. $x^{-2}/x^{-3} = x^{-2-(-3)} = x^{-2+3} = x^1 = x$

3. a. Order wrt $CH_3CHO = 3/2$, overall order $= 3/2$
   b. Order wrt $BrO_3^- = 1$, wrt $Br^- = 1$, wrt $H^+ = 2$, overall order $= 1 + 1 + 2 = 4$
   c. Order wrt $NO = 2$, wrt $Cl_2 = 1$, overall order $= 2 + 1 = 3$

4. $Rate = k[H_2O_2][H^+][Br^-] = k[H_2O_2][Br^-][Br^-] = k[H_2O_2][Br^-]^2$
   $= k[H_2O_2][H^+][H^+] = k[H_2O_2][H^+]^2$

5. Since $[H^+] = [HCO_3^-]$, $K_a = [H^+]^2/[H_2CO_3] = [HCO_3^-]^2/[H_2CO_3]$

Chapter 2

1. a. $F = 9.649 \times 10^4$ C mol^{-1} = 96 490 C mol^{-1}
   b. $R = 1.097 \times 10^7$ m^{-1} = 10 970 000 m^{-1}
   c. $\mu_0 = 12.57 \times 10^{-7}$ N A^{-2} = 0.000 001 257 N A^{-2}
   d. $a_0 = 5.292 \times 10^{-11}$ m = 0.000 000 000 052 92 m
   e. $h = 6.626 \times 10^{-34}$ J s = 0.000 000 000 000 000 000 000 000 000 000 000 662 6 J s

2. a. $e = 0.1602 \times 10^{-18}$ C $= 1.602 \times 10^{-19}$ C
   b. $E_h = 4360 \times 10^{-22}$ J $= 4.360 \times 10^{-19}$ J
   c. $m_e = 0.009\,109 \times 10^{-28}$ kg $= 9.109 \times 10^{-31}$ kg
   d. $N_A = 602.2 \times 10^{21}$ mol^{-1} $= 6.022 \times 10^{23}$ mol^{-1}
   e. $R = 0.008\,314 \times 10^4$ J K^{-1} mol^{-1} = 8.314 J K^{-1} mol^{-1}

3. a. 9.4×10^{-5} bar = 0.000 094 bar
   b. 3.72×10^{-2} cm = 0.0372 cm
   c. 1.8×10^{-4} MHz $= 1.8 \times 10^{-4} \times 10^3$ Hz $= 1.8 \times 10^{-1}$ Hz = 0.18 Hz
   d. 1.95×10^{-3} kJ mol^{-1} = 1.95 J mol^{-1}
   e. 7.19×10^4 s^{-1} = 71 900 s^{-1}

4. a. 0.0417 nm $= 0.0417 \times 10^{-9}$ m $= 4.17 \times 10^{-11}$ m
   b. 352 s $= 3.52 \times 10^2$ s
   c. 2519 m s^{-1} $= 2.519 \times 10^3$ m s^{-1}
   d. 0.076 kJ mol^{-1} $= 7.6 \times 10^{-2} \times 10^3$ J mol^{-1} $= 7.6 \times 10$ J mol^{-1}
   e. 579 eV $= 5.79 \times 10^2$ eV

Chapter 3

1. a. 54.2 mg $= 54.2 \times 10^{-3}$g $= 5.42 \times 10^{-2}$ g
   b. 1.47 aJ $= 1.47 \times 10^{-18}$ J
   c. 3.62 MW $= 3.62 \times 10^6$ W
   d. 4.18 kJ mol^{-1} $= 4.18 \times 10^3$ J mol^{-1}
   e. 589 nm $= 589 \times 10^{-9}$ m $= 5.89 \times 10^{-7}$ m

2. a. 3.0×10^8 m s^{-1} $= 0.30 \times 10^9$ m s^{-1} = 0.30 Gm s^{-1} = 300 Mm s^{-1}
   b. 101 325 Pa $= 101.325 \times 10^3$ Pa = 101.325 kPa
   c. 1.543×10^{-10} m $= 0.1543 \times 10^{-9}$ m = 0.1543 nm
   d. 1.68×10^2 kg m^{-3} $= 1.68 \times 10^5$ g m^{-3} $= 0.168 \times 10^6$ g m^{-3} = 0.168 Mg m^{-3}
   e. 7.216×10^{-4} mol dm^{-3} $= 0.7216 \times 10^{-3}$ mol dm^{-3} = 0.7216 mmol dm^{-3}

3. a. $1.082 \text{ g cm}^{-3} = 1.082\ (10^{-3}\text{ kg})\ (10^{-2}\text{ m})^{-3} = 1.082 \times 10^{-3}\text{ kg} \times 10^{6}\text{ m}^{-3} = 1.082 \times 10^{3}\text{ kg m}^{-3}$
   b. $135\text{ kPa} = 135 \times 10^{3}\text{ N m}^{-2} = 135 \times 10^{3}\text{ N}\ (10^{2}\text{ cm})^{-2} = 135 \times 10^{3}\text{ N} \times 10^{-4}\text{ cm}^{-2} = 135 \times 10^{-1}\text{ N cm}^{-2}$ $= 13.5\text{ N cm}^{-2}$
   c. $5.03\text{ mmol dm}^{-3} = 5.03 \times 10^{-3}\text{ mol}\ (10^{-1}\text{ m})^{-3} = 5.03 \times 10^{-3}\text{ mol} \times 10^{3}\text{ m}^{-3} = 5.03\text{ mol m}^{-3}$
   d. $9.81\text{ m s}^{-2} = 9.81\ (10^{2}\text{ cm})\ (10^{3}\text{ ms})^{-2} = 9.81 \times 10^{2}\text{ cm} \times 10^{-6}\text{ m s}^{-2} = 9.81 \times 10^{-4}\text{ cm ms}^{-2}$
   e. $1.47\text{ kJ mol}^{-1} = 1.47 \times 10^{3}\text{ J}\ (10^{3}\text{ mmol})^{-1} = 1.47 \times 10^{3}\text{ J} \times 10^{-3}\text{ mmol}^{-1} = 1.47\text{ J mmol}^{-1}$

4. a. $4.28\text{ Å} = 4.28 \times 10^{-10}\text{ m} = 428 \times 10^{-12}\text{ m} = 428\text{ pm}$
   b. $54.71\text{ kcal} = 54.71 \times 4.184\text{ J} = 228.9\text{ kJ}$
   c. $3.6\text{ atm} = 3.6 \times 101.325\text{ kPa} = 370\text{ kPa}$
   d. $2.91E_h = 2.91 \times 4.360 \times 10^{-18}\text{ J} = 1.27 \times 10^{-17}\text{ J}$
   e. $3.21a_0 = 3.21 \times 5.292 \times 10^{-11}\text{ m} = 1.70 \times 10^{-10}\text{ m} = 0.170 \times 10^{-9}\text{ m} = 0.170\text{ nm}$

5. a. $5.62\text{ g} \times 4.19\text{ m s}^{-2} = 23.5\text{ g m s}^{-2} = 23.5 \times 10^{-3}\text{ kg m s}^{-2} = 2.35 \times 10^{-2}\text{ kg m s}^{-2} = 2.35 \times 10^{-2}\text{ N}$
   b. $4.31\text{ kN}/10.46\text{ m}^2 = 0.412\text{ kN m}^{-2} = 412\text{ N m}^{-2} = 412\text{ Pa}$
   c. $2.118 \times 10^{-3}\text{ J}/3.119 \times 10^{-8}\text{ C} = 6.791 \times 104\text{ J C}^{-1} = 6.791 \times 10^{4}\text{ V} = 67.91 \times 10^{3}\text{ V} = 67.91\text{ kV}$
   d. $6.63 \times 10^{-34}\text{ J s} \times 3 \times 10^{8}\text{ m s}^{-1}/909\text{ nm} = 6.63 \times 10^{-34}\text{ J s} \times 3 \times 10^{8}\text{ m s}^{-1}/909 \times 10^{-9}\text{ m} = 2 \times 10^{-19}\text{ J}$
   e. $4.16 \times 10^{3}\text{ Pa} \times 2.14 \times 10^{-2}\text{ m}^3 = 8.90 \times 10\text{ N m}^{-2}\text{ m}^3 = 89.0\text{ N m} = 89.0\text{ J}$

6. a. p(Pa) should be p/Pa; T(K) should be T/K.
   b. t/sec should be t/s; c/mol dm^{-3} is correct.
   c. c/mol per dm^3 should be c/mol dm^{-3}; ρ/g cm^{-3} is correct.
   d. $\frac{1}{c/\text{mol dm}^{-3}}$ should be mol dm$^{-3}/c$; t/s is correct.
   e. n_i/n_j is correct; $T/\text{K}(\times 10^3)$ should be $T/10^{-3}$ K or $T/10^3$ K, i.e. the expression is ambiguous.

Chapter 4

1. a. (i) 41.6 (ii) 41.62
   b. (i) 3.96 (ii) 3.96
   c. (i) 10 000, since internal zeros are significant (ii) 10 004.91
   d. (i) 0.007 16, since leading zeros are not significant (ii) 0.01
   e. (i) 1.00 (ii) 1.00

2. a. (i) 589.9 nm (ii) 589.9 nm
   b. (i) 103.1 kJ (ii) 103.1 kJ
   c. (i) 0.100 5 mol dm^{-3} (ii) 0.1 mol dm^{-3}
   d. (i) 32.85 ms (ii) 32.8 ms
   e. (i) 101 300 Pa (ii) 101 325.0 Pa

3. a. 4 sf
   b. 5 sf, since internal zeros are significant
   c. 5 sf, since leading zeros are not significant
   d. 4 sf, since trailing zeros after the decimal point are significant
   e. 6 sf, since internal zeros are significant and trailing zeros after the decimal point are significant

4. a. 4 sf
   b. 4 sf, since internal zeros are significant
   c. 3 sf, since leading zeros are not significant
   d. 4 sf, since trailing zeros after the decimal point are significant
   e. 4 sf, since leading zeros are not significant and internal zeros are significant

Chapter 5

1. a. $1.092 + 2.43 = 3.52$, since 2.43 is given to 2 dp
   b. $6.2468 - 1.3 = 4.9$, since 1.3 is given to 1 dp
   c. $100 + 9.1 = 109$, since 100 is given to 0 dp
   d. $42.8 \times 36.194 = 1550$, since 42.8 is given to 3 sf
   e. $2.107/32 = 0.066$, since 32 is given to 2 sf

2. a. 9.021 g/10.7 cm^3 = 0.843 g cm^{-3}, since 10.7 cm^3 is given to 3 sf
   b. 104.6 kJ mol^{-1} + 98.14 kJ mol^{-1} = 202.7 kJ mol^{-1}, since 104.6 kJ mol^{-1} is given to 1 dp
   c. 1.46 mol/12.2994 dm^3 = 0.119 mol dm^{-3}, since 1.46 mol is given to 3 sf
   d. 3.61 kg × 2.1472 m s^{-1} = 7.75 kg m s^{-1}, since 3.61 kg is given to 3 sf
   e. 3.2976 g – 0.004 g = 3.294 g, since 0.004 g is given to 3 dp

3. a. $V = \dfrac{2.42\,\text{mol} \times 8.314\,\text{J K mol}^{-1} \times 295\,\text{K}}{52.47 \times 10^3\,\text{N m}^{-2}} = 0.113\,\text{m}^3$
   b. V = 0.11 m^3, since R is given to 2 sf
   c. V = 0.113 m^3, since R is given to 3 sf
   d. V = 0.113 m^3, since n and T are given to 3 sf

4. a. m(Al) = 2 × 26.981 153 9 g = 53.963 078 g, since 2 is an exact integer; m(HCl) = 6 × (1.007 94 + 35.4527) g = 6 × 36.4606 g = 218.764 g, since 35.4527 g is given to 4 dp and 36.4606 g is given to 6 sf
   b. Mass of HCl unreacted = 300 g – 218.764 g = 81 g, since 300 g is given to 0 dp
   c. Mass of Al unreacted = 100 g – 53.963 078 g = 46 g, since 100 g is given to 0 dp

Chapter 6

1. a. Absolute error = 16.72 – 16.87 = –0.15
   b. Fractional error = –0.15/16.87 = –8.9 × 10^{-3}
   c. Percentage error = –8.9 × 10^{-3} × 100 = –0.89%

2. a. Absolute error = 482 nm – 472 nm = 10 nm
   b. Fractional error = 10 nm/472 nm = 0.021
   c. Percentage error = 0.021 × 100 = 2.1%

3. a. Absolute error = ±0.05 V, as given
   b. Fractional error = ±0.05 V/6.45 V = ±8 × 10^{-3}
   c. Percentage error = ±8 × 10^{-3} × 100 = ±0.8%

4. a. 2.3° – (–0.1°) = 2.4°
   b. 4.6° – (–0.1°) = 4.7°
   c. 9.8° – (–0.1°) = 9.9°

5. a. True time = 37.0 s – 0.5 s = 36.5 s
   b. Absolute error = ±0.5 s, as given
   c. Fractional error = ±0.5 s/36.5 s = ±0.01
   d. Percentage error = ±0.01 × 100 = ±1%

Chapter 7

1. Maximum value of $\Delta_{sub}H$ = (8.3 + 0.1) kJ mol^{-1} + (16.9 + 0.2) kJ mol^{-1} = (8.4 + 17.1) kJ mol^{-1} = 25.5 kJ mol^{-1}
   Minimum value of $\Delta_{sub}H$ = (8.3 – 0.1) kJ mol^{-1} + (16.9 – 0.2) kJ mol^{-1} = (8.2 + 16.7) kJ mol^{-1} = 24.9 kJ mol^{-1}
   Maximum possible error in $\Delta_{sub}H$ = ± (25.5 – 24.9) kJ mol^{-1}/2 = ±0.3 kJ mol^{-1}

2. Maximum value of EMF = (0.76 + 0.01) V + (0.34 + 0.01) V = 0.77 V + 0.35 V = 1.12 V
   Minimum value of EMF = (0.76 – 0.01) V + (0.34 – 0.01) V = 0.75 V + 0.33 V = 1.08 V
   Maximum possible error in overall EMF = ±(1.12 – 1.08) V/2 = ±0.02 V

3. Maximum value of K_s = 4 × [(1.62 + 0.02) × 10^{-2} mol dm^{-3}]^3 = 4 × (1.64 × 10^{-2} mol dm^{-3})^3 = 4 × 4.41 × 10^{-6} mol^3 dm^{-9} = 1.76 × 10^{-5} mol^3 dm^{-9}
   Minimum value of K_s = 4 × [(1.62 – 0.02) × 10^{-2} mol dm^{-3}]^3 = 4 × (1.60 × 10^{-2} mol dm^{-3})^3 = 4 × 4.10 × 10^{-6} mol^3 dm^{-9} = 1.64 × 10^{-5} mol^3 dm^{-9}
   Maximum possible error in K_s = ±(1.76 – 1.64) × 10^{-5} mol^3 dm^{-9}/2 = ±6.2 × 10^{-7} mol^3 dm^{-9}

4. Maximum value of ρ = (521 + 5) g/(27 – 1) cm^3 = 526 g/26 cm^3 = 20 g cm^{-3}
   Minimum value of ρ = (521 – 5) g/(27 + 1) cm^3 = 516 g/28 cm^3 = 18 g cm^{-3}
   Maximum possible error in ρ = ± (20 – 18) g cm^{-3}/2 = ±1 g cm^{-3}

5. Maximum value of $\Delta G^{\ominus} = (-58.0 + 0.5) \times 10^3$ J mol^{-1} – (298 –1) K × (−177 –1) J K^{-1} mol^{-1} = −57 500 J mol^{-1} + 52 866 J mol^{-1} = −4634 J mol^{-1}
   Minimum value of $\Delta G^{\ominus} = (-58.0 - 0.5) \times 10^3$ J mol^{-1} – (298 + 1) K × (−177 + 1) J K^{-1} mol^{-1} = −58 500 J mol^{-1} + 52 624 J mol^{-1} = −5 876 J mol^{-1}
   Maximum possible error in $\Delta G^{\ominus} = \pm\,(-4634 - (-5876))$ J mol^{-1}/2 = ±621 J mol^{-1}

Chapter 8

1. Maximum probable error in $\Delta S = \sqrt{[(0.004\,\text{J K}^{-1}\,\text{mol}^{-1})^2 + (0.1\,\text{J K}^{-1}\,\text{mol}^{-1})^2 + (0.2\,\text{J K}^{-1}\,\text{mol}^{-1})^2]} = \sqrt{(1.6 \times 10^{-5} + 0.01 + 0.04)}\,\text{J K}^{-1}\,\text{mol}^{-1} = \sqrt{0.05}\,\text{J K}^{-1}\,\text{mol}^{-1} = 0.22\,\text{J K}^{-1}\,\text{mol}^{-1}$, since the first term in square root can be neglected.

2. Maximum probable error in wavelengths $= \sqrt{[(0.007\,\text{nm})^2 + (0.005\,\text{nm})^2]} = \sqrt{[(4.9 \times 10^{-5}\,\text{nm}^2) + (2.5 \times 10^{-5}\,\text{nm}^2)]} = \sqrt{(7.4 \times 10^{-5}\,\text{nm}^2)} = 8.6 \times 10^{-3}\,\text{nm} = 0.009\,\text{nm}$

3. $Rate = 9.3 \times 10^{-5}\,\text{s}^{-1} \times 0.105\,\text{mol dm}^{-3} = 9.8 \times 10^{-6}\,\text{mol dm}^{-3}\,\text{s}^{-1}$

$$\frac{\text{Maximum probable error in rate}}{9.8 \times 10^{-6}\,\text{mol dm}^{-3}\,\text{s}^{-1}} = \sqrt{\left[\left(\frac{0.1}{9.3}\right)^2 + \left(\frac{0.003}{0.105}\right)^2\right]} = \sqrt{[(1.16 \times 10^{-4}) + (8.16 \times 10^{-4})]}$$
$$= \sqrt{(9.32 \times 10^{-4})} = 0.0305$$

   Maximum probable error in rate = 9.8×10^{-6} mol dm^{-3} s^{-1} × 0.0305 = 3 × 10^{-7} mol dm^{-3} s^{-1}

4. $\Delta_{vap}S = 29.4 \times 10^3\,\text{J mol}^{-1}/334\,\text{K} = 88.0\,\text{J K}^{-1}\,\text{mol}^{-1}$

$$\frac{\text{Maximum probable error in}\,\Delta_{vap}S}{88.0\,\text{J K}^{-1}\,\text{mol}^{-1}} = \sqrt{\left[\left(\frac{0.1}{29.4}\right)^2 + \left(\frac{1}{334}\right)^2\right]} = \sqrt{[(1.16 \times 10^{-5}) + (8.96 \times 10^{-6})]}$$
$$= \sqrt{(2.06 \times 10^{-5})} = 4.54 \times 10^{-3}$$

   Maximum probable error in $\Delta_{vap}S = 4.54 \times 10^{-3} \times 88.0\,\text{J K}^{-1}\,\text{mol}^{-1} = 0.4\,\text{J K}^{-1}\,\text{mol}^{-1}$

5. $RT\Delta n = 8.311\,\text{J K}^{-1}\,\text{mol}^{-1} \times 298\,\text{K} \times (-1\,\text{mol}) = -2\,476\,\text{J mol}^{-1}$

$$\frac{\text{Maximum probable error in}\,\Delta H}{2476\,\text{J mol}^{-1}} = \sqrt{\left[\left(\frac{0.05}{8.31}\right)^2 + \left(\frac{1}{298}\right)^2\right]} = \sqrt{[(3.62 \times 10^{-5}) + (1.13 \times 10^{-5})]}$$
$$= \sqrt{(4.75 \times 10^{-5})} = 6.89 \times 10^{-3}$$

   Maximum probable error in $\Delta H = 6.89 \times 10^{-3} \times 2476\,\text{J mol}^{-1} = 17.1\,\text{J mol}^{-1}$

$$\begin{aligned}\text{Absolute error in}\,\Delta H &= \sqrt{[(100\,\text{J mol}^{-1})^2 + (17\,\text{J mol}^{-1})^2]}\\ &= \sqrt{[10\,000\,\text{J}^2\,\text{mol}^{-2} + 290\,\text{J}^2\,\text{mol}^{-2}]}\\ &= \sqrt{(10\,290\,\text{J}^2\,\text{mol}^{-2})}\\ &= 101\,\text{J mol}^{-1}\end{aligned}$$

Chapter 9

1. $\bar{c} = 1.934\,\mu\text{g m}^{-3}/6 = 0.322\,\mu\text{g m}^{-3}$, $s^2 = 1.42 \times 10^{-2}\,\mu\text{g}^2\,\text{m}^{-6}/(6-1) = 2.84 \times 10^{-3}\,\mu\text{g}^2\,\text{m}^{-6}$, $s = 0.053\,\mu\text{g m}^{-3}$

2. $\bar{c} = 7.276\,\text{mg dm}^{-3}/8 = 0.910\,\text{mg dm}^{-3}$, $s^2 = 0.893\,\text{mg}^2\,\text{dm}^{-6}/(8-1) = 0.128\,\text{mg}^2\,\text{dm}^{-6}$, $s = 0.358\,\text{mg dm}^{-3}$

3. $\bar{p} = 5075.85\,\text{kPa}/5 = 1015.17\,\text{kPa}$, $s^2 = 1.7404\,\text{kPa}^2/(5-1) = 0.44\,\text{kPa}^2$, $s = 0.66\,\text{kPa}$

4. $\bar{l} = 14.342\,\text{Å}/7 = 2.049\,\text{Å}$, $s^2 = 0.0695\,\text{Å}^2/(7-1) = 0.0116\,\text{Å}^2$, $s = 0.108\,\text{Å}$

Chapter 10

1. $s = 0.032$ Å, $n = 7$, $n - 1 = 6$, $t = 1.94$

$$\frac{ts}{\sqrt{n}} = \frac{1.94 \times 0.032\,\text{Å}}{\sqrt{7}} = 0.023\,\text{Å}$$

2. $s = 0.88$ ppb, $n = 5$, $n - 1 = 4$, $t = 2.78$

$$\frac{ts}{\sqrt{n}} = \frac{2.78 \times 0.88\,\text{ppb}}{\sqrt{5}} = 1.09\,\text{ppb}$$

3. $s = 0.22$, $n = 8$, $n - 1 = 7$, $t = 2.36$

$$\frac{ts}{\sqrt{n}} = \frac{2.36 \times 0.22}{\sqrt{8}} = 0.18$$

4. $s = 0.008\ \text{mg dm}^{-3}$, $n = 6$, $n - 1 = 5$, $t = 3.16$

$$\frac{ts}{\sqrt{n}} = \frac{3.16 \times 0.008\,\text{mg dm}^{-3}}{\sqrt{6}} = 0.010\,\text{mg dm}^{-3}$$

5. $s = 0.56$ kPa, $n = 5$, $n - 1 = 4$, $t = 4.60$

$$\frac{ts}{\sqrt{n}} = \frac{4.60 \times 0.56\,\text{kPa}}{\sqrt{5}} = 1.15\,\text{kPa}$$

Chapter 11

1. $$K = \frac{16 \times 0.15^2(1 - 0.15)}{(1 - 3 \times 0.15)^3 \left(\dfrac{2.5\,\text{atm}}{1\,\text{atm}}\right)^2} = \frac{16 \times 0.0225 \times 0.85}{(1 - 0.45)^3 \times 2.5^2} = \frac{0.306}{0.55^3 \times 2.5^2}$$
$$= \frac{0.306}{0.1664 \times 6.25} = \frac{0.306}{1.040} = 0.29$$

2. $$\frac{\Delta H^\ominus}{R}\left(\frac{1}{T_1} - \frac{1}{T_2}\right) = \frac{38\,400\,\text{J mol}^{-1}}{8.314\,\text{J K}^{-1}\,\text{mol}^{-1}}\left(\frac{1}{298\,\text{K}} - \frac{1}{300\,\text{K}}\right)$$
$$= 4619\,\text{K}(3.356 \times 10^{-3}\,\text{K}^{-1} - 3.333 \times 10^{-3}\,\text{K}^{-1}) = 4619\,\text{K} \times 0.023 \times 10^{-3}\,\text{K}^{-1}$$
$$= 0.106$$

3. $$p = \frac{3125\,\text{Pa} \times 2967\,\text{Pa}}{3125\,\text{Pa} + (2967\,\text{Pa} - 3125\,\text{Pa}) \times 0.365} = \frac{9271875\,\text{Pa}^2}{3125\,\text{Pa} + (-158\,\text{Pa}) \times 0.365}$$
$$= \frac{9\,271\,875\,\text{Pa}^2}{3125\,\text{Pa} - 57.67\,\text{Pa}} = \frac{9\,271\,875\,\text{Pa}^2}{3067\,\text{Pa}} = 3023\,\text{Pa}$$

4. $V/\text{cm}^3 = 18.023 + 53.57 \times 0.27 + 1.45 \times 0.27^2 = 18.023 + 53.57 \times 0.27 + 1.45 \times 0.0729 = 18.023 + 14 + 0.106 = 32$

Chapter 12

1. a. $\frac{1}{3} + \frac{1}{6} = \frac{2}{6} + \frac{1}{6} = \frac{3}{6} = \frac{1}{2}$

   b. $\frac{3}{4} + \frac{2}{3} = \frac{9 + 8}{12} = \frac{17}{12}$

   c. $\frac{2}{3} + \frac{1}{8} = \frac{16 + 3}{24} = \frac{19}{24}$

   d. $\frac{2}{3} - \frac{1}{4} = \frac{8 - 3}{12} = \frac{5}{12}$

   e. $\frac{4}{3} - \frac{3}{16} = \frac{64 - 9}{48} = \frac{55}{48}$

2. a. $\frac{1}{2} \times \frac{3}{4} = \frac{3}{8}$
   b. $\frac{3}{8} \times \frac{3}{4} = \frac{9}{32}$
   c. $\frac{1}{4} \times \frac{22}{7} = \frac{22}{28} = \frac{11}{14}$
   d. $\frac{2}{3} \div \frac{3}{16} = \frac{2}{3} \times \frac{16}{3} = \frac{32}{9}$
   e. $\frac{1}{2} \div \frac{3}{4} = \frac{1}{2} \times \frac{4}{3} = \frac{4}{6} = \frac{2}{3}$

3. $n_S = \dfrac{33.4\,\text{g}}{32.1\,\text{g mol}^{-1}} = 1.04\,\text{mol},\ n_O = \dfrac{50.1\,\text{g}}{16.0\,\text{g mol}^{-1}} = 3.13\,\text{mol}$

$$\frac{n_O}{n_S} = \frac{3.13\,\text{mol}}{1.04\,\text{mol}} = 3.0$$

   Empirical formula is SO_3

4. $\tilde{\nu} = R\left(\frac{1}{m^2} - \frac{1}{n^2}\right) = R\left(\frac{1}{1^2} - \frac{1}{3^2}\right) = R\left(1 - \frac{1}{9}\right) = \frac{R}{9}(9-1)$
   $= \frac{8R}{9} = 9.75 \times 10^4\,\text{cm}^{-1}$

5. $\tilde{\nu} = R\left(\frac{1}{m^2} - \frac{1}{n^2}\right) = R\left(\frac{1}{1} - \frac{1}{\infty^2}\right) = R(1-0) = R = 1.097 \times 10^5\,\text{cm}^{-1}$

Chapter 13

1. a. $x + 4 > 13, x + 4 - 4 > 13 - 4, x > 9$
   b. $y + 7 > 25, y + 7 - 7 > 25 - 7, y > 18$
   c. $x - 3 < 10, x - 3 + 3 < 10 + 3, x < 13$
   d. $x - 10 \le 16, x - 10 + 10 \le 16 + 10, x \le 26$
   e. $5 + y \ge 17, 5 + y - 5 \ge 17 - 5, y \ge 12$

2. a. $9 - x > 2, 9 - x - 9 > 2 - 9, -x > -7, x < 7$
   b. $4 - x < 3, 4 - x - 4 < 3 - 4, -x < -1, x > 1$
   c. $2 - y < -6, 2 - y - 2 < -6 - 2, -y < -8, y > 8$
   d. $14 - x \ge 7, 14 - x - 14 \ge 7 - 14, -x \ge -7, x \le 7$
   e. $6 - y \le 1, 6 - y - 6 \le 1 - 6, -y \le -5, y \ge 5$

3. a. $3x > 18, x > 18/3, x > 6$
   b. $4x + 2 < 18, 4x + 2 - 2 < 18 - 2, 4x < 16, x < 16/4, x < 4$
   c. $9 - 3x \ge 72, 9 - 3x - 9 \ge 72 - 9, -3x \ge 63, 3x \le -63, x \le -63/3, x \le -21$
   d. $3y - 7 \le 28, 3y - 7 + 7 \le 28 + 7, 3y \le 35, y \le 35/3$
   e. $5 - 4x \le 29, 5 - 4x - 5 \le 29 - 5, -4x \le 24, 4x \ge -24, x \ge -24/4, x \ge -6$

4. Since $E_1 > E_2$, according to the variation theorem Ψ_2 must be the true wavefunction.

5. $\Delta G = \Delta H - T\Delta S = -91\,800\ \text{J mol}^{-1} - 298\ \text{K} \times (-197\ \text{J K}^{-1}\ \text{mol}^{-1}) = -91\,800\ \text{J mol}^{-1} + 58\,706\ \text{J mol}^{-1} = -33\,094\ \text{J mol}^{-1}$. $\Delta G < 0$, so reaction is spontaneous.

Chapter 14

1. a. $x + y = 4, x + y - x = 4 - x, y = 4 - x$
   b. $3x + 2y = 17, 3x + 2y - 3x = 17 - 3x, 2y = 17 - 3x, \frac{2y}{2} = \frac{17-3x}{2}, y = \frac{17-3x}{2}$
   c. $x^2 - y^2 = 5, x^2 - y^2 + y^2 = 5 + y^2, x^2 = 5 + y^2, x^2 - 5 = 5 + y^2 - 5, y^2 = x^2 - 5, y = \surd(x^2 - 5)$
   d. $4x^2y = 20, \frac{4x^2y}{4x^2} = \frac{20}{4x^2}, y = \frac{5}{x^2}$
   e. $3x^2y^2 + 2 = 19, 3x^2y^2 + 2 - 2 = 19 - 2, 3x^2y^2 = 17, \frac{3x^2y^2}{3x^2} = \frac{17}{3x^2}, y^2 = \frac{17}{3x^2}, y = \surd(\frac{17}{3x^2}) = \frac{1}{x}\surd(\frac{17}{3})$

2. $\frac{pV}{V} = \frac{nRT}{V}, p = \frac{nRT}{V}; \frac{pV}{p} = \frac{nRT}{p}, V = \frac{nRT}{p}; \frac{pV}{RT} = \frac{nRT}{RT}, n = \frac{pV}{RT}; \frac{pV}{nR} = \frac{nRT}{nR}, T = \frac{pV}{nR}$

3. $P + F = C - 2, P + F - P = C - 2 - P, F = C - P - 2$

4. $\Delta G = \Delta H - T\Delta S, \Delta G + T\Delta S = \Delta H - T\Delta S + T\Delta S$
   $\Delta G + T\Delta S = \Delta H, \Delta G + T\Delta S - \Delta G = \Delta H - \Delta G, T\Delta S = \Delta H - \Delta G$

$$\frac{T\Delta S}{T} = \frac{\Delta H - \Delta G}{T} \qquad \Delta S = \frac{\Delta H - \Delta G}{T}$$

   When $\Delta G = 0$, $\Delta S = \frac{\Delta H}{T}$

5. $k_1[C_2H_6] - k_2[CH_3][C_2H_6] = 0$
   $k_1[C_2H_6] - k_2[CH_3][C_2H_6] + k_2[CH_3][C_2H_6] = 0 + k_2[CH_3][C_2H_6]$
   $k_1[C_2H_6] = k_2[CH_3][C_2H_6]$

$$\frac{k_1[C_2H_6]}{k_2[C_2H_6]} = \frac{k_2[CH_3][C_2H_6]}{k_2[C_2H_6]} \qquad [CH_3] = \frac{k_1}{k_2}$$

6. $\tilde{\nu} = R\left(\frac{1}{n_1^2} - \frac{1}{n_2^2}\right), \frac{\tilde{\nu}}{R} = \frac{R}{R}\left(\frac{1}{n_1^2} - \frac{1}{n_2^2}\right), \frac{\tilde{\nu}}{R} = \frac{1}{n_1^2} - \frac{1}{n_2^2},$

$$\frac{\tilde{\nu}}{R} + \frac{1}{n_2^2} = \frac{1}{n_1^2} - \frac{1}{n_2^2} + \frac{1}{n_2^2}, \frac{\tilde{\nu}}{R} + \frac{1}{n_2^2} = \frac{1}{n_1^2}, \frac{\tilde{\nu}}{R} + \frac{1}{n_2^2} - \frac{\tilde{\nu}}{R} = \frac{1}{n_1^2} - \frac{\tilde{\nu}}{R}, \frac{1}{n_2^2} = \frac{1}{n_1^2} - \frac{\tilde{\nu}}{R},$$

$$\frac{1}{n_2^2} = \frac{R - \tilde{\nu}n_1^2}{Rn_1^2}, n_2^2 = \frac{Rn_1^2}{R - \tilde{\nu}n_1^2}$$

$$n_2 = \sqrt{\frac{Rn_1^2}{R - \tilde{\nu}n_1^2}} = n_1\sqrt{\frac{R}{R - \tilde{\nu}n_1^2}}$$

Chapter 15

1. $y/x = 4$, $y = 4x$, direct proportionality

2. $xy = 5$, $y = 5/x$, inverse proportionality

3. xy varies; $xz = 1$, so $z = 1/x$, inverse proportionality; $yz = 3$, so $z = 3/y$, inverse proportionality; $y/x = 3$, so $y = 3x$, direct proportionality; x/z varies, y/z varies

4. $F = kx$, $1.6 \times 10^{-9}\,\mathrm{N} = k \times 5 \times 10^{-12}\,\mathrm{m}$

$$k = \frac{1.6 \times 10^{-9}\,\mathrm{N}}{5 \times 10^{-12}\,\mathrm{m}} = 320\,\mathrm{N\,m^{-1}}$$

5. $\lambda = \frac{k}{\nu}$, $510\,\mathrm{nm} = \frac{k}{5.88 \times 10^{14}\,\mathrm{Hz}}$
   $k = 510\,\mathrm{nm} \times 5.88 \times 10^{14}\,\mathrm{Hz} = 510 \times 10^{-9}\,\mathrm{m} \times 5.88 \times 10^{14}\,\mathrm{s^{-1}} = 3.00 \times 10^8\,\mathrm{m\,s^{-1}}$

6. $rate = kc$, $k = rate/c$

$$k = \frac{0.92\,\mathrm{s^{-1}}}{0.1\,\mathrm{mol\,dm^{-3}}} = \frac{0.46\,\mathrm{s^{-1}}}{0.05\,\mathrm{mol\,dm^{-3}}} = 9.2\,\mathrm{mol\,dm^{-3}\,s^{-1}}$$

Chapter 16

1. a. $3! = 3 \times 2 \times 1 = 6$
   b. $4! = 4 \times 3 \times 2 \times 1 = 24$
   c. $5! = 5 \times 4 \times 3 \times 2 \times 1 = 120$
   d. $6! = 6 \times 5 \times 4 \times 3 \times 2 \times 1 = 720$
   e. $7! = 7 \times 6 \times 5 \times 4 \times 3 \times 2 \times 1 = 5040$

2. a. $10! = 3\,628\,800$
   b. $12! = 479\,001\,600$
   c. $15! = 1.308 \times 10^{12}$
   d. $25! = 1.551 \times 10^{25}$
   e. $36! = 3.720 \times 10^{41}$

3. a. $\frac{8!}{6!} = \frac{8 \times 7 \times 6!}{6!} = 8 \times 7 = 56$
   b. $\frac{5!}{4!} = \frac{5 \times 4!}{4!} = 5$
   c. $\frac{20!}{17!} = \frac{20 \times 19 \times 18 \times 17!}{17!} = 20 \times 19 \times 18 = 6840$
   d. $\frac{10!}{8!} \times \frac{5!}{7!} = \frac{10 \times 9 \times 8!}{8!} \times \frac{5!}{7 \times 6 \times 5!} = \frac{10 \times 9}{7 \times 6} = \frac{90}{42} = \frac{15}{7}$
   e. $\frac{36!}{30!} = \frac{36 \times 35 \times 34 \times 33 \times 32 \times 31 \times 30!}{30!} = 36 \times 35 \times 34 \times 33 \times 32 \times 31 = 1.402 \times 10^9$

4. a. ${}^4C_2 = \frac{4!}{(4-2)!\,2!} = \frac{4!}{2!\,2!} = \frac{4 \times 3 \times 2!}{2!\,2!} = \frac{4 \times 3}{2!} = \frac{4 \times 3}{2 \times 1} = \frac{12}{2} = 6$
   b. ${}^5C_3 = \frac{5!}{(5-3)!\,3!} = \frac{5!}{2!\,3!} = \frac{5 \times 4 \times 3!}{2!\,3!} = \frac{5 \times 4}{2!} = \frac{5 \times 4}{2 \times 1} = \frac{20}{2} = 10$
   c. ${}^5C_4 = \frac{5!}{(5-4)!\,4!} = \frac{5!}{1!\,4!} = \frac{5 \times 4!}{1!\,4!} = \frac{5}{1!} = \frac{5}{1} = 5$
   d. ${}^8C_6 = \frac{8!}{(8-6)!\,6!} = \frac{8!}{2!\,6!} = \frac{8 \times 7 \times 6!}{2!\,6!} = \frac{8 \times 7}{2!} = \frac{8 \times 7}{2 \times 1} = \frac{56}{2} = 28$
   e. ${}^8C_7 = \frac{8!}{(8-7)!\,7!} = \frac{8!}{1!\,7!} = \frac{8 \times 7!}{1!\,7!} = \frac{8}{1!} = \frac{8}{1} = 8$

5. CH_3 peaks are split into (2 + 1) by CH_2, i.e. 3 peaks of relative intensity ${}^2C_0$, ${}^2C_1$ and ${}^2C_2$:
   ${}^2C_0 = \frac{2!}{(2-0)!\,0!} = \frac{2!}{2!\,0!} = \frac{1}{0!} = \frac{1}{1} = 1$
   ${}^2C_1 = \frac{2!}{(2-1)!\,1!} = \frac{2!}{1!\,1!} = \frac{2 \times 1}{1 \times 1} = 2$
   ${}^2C_2 = \frac{2!}{(2-2)!\,2!} = \frac{2!}{0!\,2!} = \frac{1}{0!} = \frac{1}{1} = 1$
   CH_2 peaks are split into (6 + 1) by 2 × CH_3 groups, i.e. 7 peaks of relative intensity ${}^6C_0$, ${}^6C_1$, ${}^6C_2$, ${}^6C_3$, ${}^6C_4$, ${}^6C_5$ and ${}^6C_6$:

$${}^6C_0 = \frac{6!}{(6-0)!\,0!} = \frac{6!}{6!\,0!} = \frac{1}{0!} = \frac{1}{1} = 1$$
$${}^6C_1 = \frac{6!}{(6-1)!\,1!} = \frac{6!}{5!\,1!} = \frac{6 \times 5!}{5!\,1!} = \frac{6}{1!} = \frac{6}{1} = 6$$
$${}^6C_2 = \frac{6!}{(6-2)!\,2!} = \frac{6!}{4!\,2!} = \frac{6 \times 5 \times 4!}{4!\,2!} = \frac{6 \times 5}{2!} = \frac{6 \times 5}{2 \times 1} = \frac{30}{2} = 15$$
$${}^6C_3 = \frac{6!}{(6-3)!\,3!} = \frac{6!}{3!\,3!} = \frac{6 \times 5 \times 4 \times 3!}{3!\,3!} = \frac{6 \times 5 \times 4}{3!} = \frac{6 \times 5 \times 4}{3 \times 2 \times 1} = \frac{120}{6} = 20$$
$${}^6C_4 = \frac{6!}{(6-4)!\,4!} = \frac{6!}{2!\,4!} = 15$$
$${}^6C_5 = \frac{6!}{(6-5)!\,5!} = \frac{6!}{1!\,5!} = 6$$
$${}^6C_6 = \frac{6!}{(6-6)!\,6!} = \frac{6!}{0!\,6!} = 1$$

6. $\Omega = \frac{10!}{4!\,3!\,2!\,1!} = \frac{10 \times 9 \times 8 \times 7 \times 6 \times 5 \times 4!}{4! \times (3 \times 2 \times 1) \times (2 \times 1) \times 1} = \frac{151\,200}{12} = 12\,600$

Chapter 17

1. a. $f(-2) = 3 \times (-2) - 4 = -6 - 4 = -10$
   b. $f(0) = 3 \times 0 - 4 = -4$
   c. $f(3) = 3 \times 3 - 4 = 9 - 4 = 5$

2. a. $f(-3) = 4 \times (-3)^2 - 2 \times (-3) - 6 = 4 \times 9 + 6 - 6 = 36$
   b. $f(0) = 4 \times 0^2 - 2 \times 0 - 6 = -6$

c. $f(2) = 4 \times 2^2 - 2 \times 2 - 6 = 4 \times 4 - 4 - 6 = 16 - 4 - 6 = 6$
d. $f(1/2) = 4 \times (1/2)^2 - 2 \times (1/2) - 6 = 4 \times (1/4) - 2 \times (1/2) - 6 = 1 - 1 - 6 = -6$

3. a. $g(-2) = \frac{1}{-2} + \frac{2}{(-2)^2} + \frac{3}{(-2)^3} = -\frac{1}{2} + \frac{2}{4} - \frac{3}{8} = \frac{-4+4-3}{8} = -\frac{3}{8}$
b. $g\left(\frac{1}{4}\right) = \frac{1}{\left(1/4\right)} + \frac{2}{\left(1/4\right)^2} + \frac{3}{\left(1/4\right)^3} = 4 + \frac{2}{\left(1/16\right)} + \frac{3}{\left(1/64\right)} = 4 + 2 \times \frac{16}{1} + 3 \times \frac{64}{1} = 4 + 32 + 192 = 228$
c. $g(4) = \frac{1}{4} + \frac{2}{4^2} + \frac{3}{4^3} = \frac{1}{4} + \frac{2}{16} + \frac{3}{64} = \frac{16+8+3}{64} = \frac{27}{64}$
$g(y)$ is undefined when $y = 0$.

4. $\rho(0.15) = 0.987 - 0.269 \times 0.15 + 0.304 \times 0.15^2 - 0.598 \times 0.15^3$
$= 0.987 - 0.269 \times 0.15 + 0.304 \times 0.0225 - 0.598 \times 0.003\,375$
$= 0.987 - 0.040\,35 + 0.006\,84 - 0.002\,018 = 0.95$

5. $V(350\,\text{pm}) = 4 \times 1.63 \times 10^{-21}\,\text{J} \times \left[\left(\frac{358\,\text{pm}}{350\,\text{pm}}\right)^{12} - \left(\frac{358\,\text{pm}}{350\,\text{pm}}\right)^{6}\right]$
$= 6.52 \times 10^{-21}\,\text{J} \times \left(1.02^{12} - 1.02^{6}\right) = 6.52 \times 10^{-21}\,\text{J} \times \left(1.3122 - 1.1455\right)$
$= 6.52 \times 10^{-21}\,\text{J} \times 0.1667 = 1.08 \times 10^{-21}\,\text{J}$

6. $p(0.3) = 0.3 \times 1.800 \times 10^5\,\text{Pa} + (1 - 0.3) \times 0.742 \times 10^5\,\text{Pa} = 5.4 \times 10^4\,\text{Pa} + 0.7 \times 0.742 \times 10^5\,\text{Pa} = 5.4 \times 10^4\,\text{Pa} + 5.19 \times 10^4\,\text{Pa} = 10.59 \times 10^4\,\text{Pa} = 1.059 \times 10^5\,\text{Pa}$

Chapter 18

1. a. $f(1, 1) = 1 + 2 \times 1 - 3 \times 1 = 1 + 2 - 3 = 0$
b. $f(0, 0) = 1 + 2 \times 0 - 3 \times 0 = 1 + 0 - 0 = 1$
c. $f(-2, 0) = 1 + 2 \times (-2) + 3 \times 0 = 1 - 4 + 0 = -3$
d. $f(-3, -2) = 1 + 2 \times (-3) - 3 \times (-2) = 1 - 6 + 6 = 1$
e. $f(0, 3) = 1 + 2 \times 0 - 3 \times 3 = 1 + 0 - 9 = -8$

2. a. $g(1, 0, -1) = 3 \times 1^2 - 4 \times 0 + (-1) = 3 - 0 - 1 = 2$
b. $g(2, 2, 0) = 3 \times 2^2 - 4 \times 2 + 0 = 3 \times 4 - 4 \times 2 + 0 = 12 - 8 = 4$
c. $g(-3, -2, 1) = 3 \times (-3)^2 - 4 \times (-2) + 1 = 3 \times 9 + 8 + 1 = 27 + 8 + 1 = 36$
d. $g(-2, 0, 3) = 3 \times (-2)^2 - 4 \times 0 + 3 = 3 \times 4 - 0 + 3 = 12 - 0 + 3 = 15$
e. $g(-3, 4, -2) = 3 \times (-3)^2 - 4 \times 4 + (-2) = 3 \times 9 - 4 \times 4 - 2 = 27 - 16 - 2 = 9$

3. a. $f(2, 1) = 2 \times 2^2 \times 1 - 3 \times 2 \times 12 = 2 \times 4 \times 1 - 3 \times 2 \times 1 = 8 - 6 = 2$
b. $f(0, 3) = 2 \times 0^2 \times 3 - 3 \times 0 \times 32 = 0$
c. $f(-2, 1) = 2 \times (-2)^2 \times 1 - 3 \times (-2) \times 12 = 2 \times 4 \times 1 + 3 \times 2 \times 1 = 8 + 6 = 14$
d. $f(1, -2) = 2 \times 1^2 \times (-2) - 3 \times 1 \times (-2)^2 = -2 \times 1 \times 2 - 3 \times 1 \times 4 = -4 - 12 = -16$
e. $f(-1, -2) = 2 \times (-1)^2 \times (-2) - 3 \times (-1) \times (-2)^2 = -2 \times 1 \times 2 + 3 \times 1 \times 4 = -4 + 12 = 8$

4. $p(300\,\text{K}, 1.5\,\text{m}^3) = \frac{2.5\,\text{mol} \times 8.314\,\text{J}\,\text{K}^{-1}\,\text{mol}^{-1} \times 298\,\text{K}}{1.5\,\text{m}^3} = 4129\,\text{N}\,\text{m}\,\text{m}^{-3}$
$= 4129\,\text{N}\,\text{m}^{-2} = 4.1 \times 10^3\,\text{Pa}$

5. $E(1, 2, 1) = \frac{(6.63 \times 10^{-34}\,\text{J}\,\text{s})^2}{8 \times 9.11 \times 10^{-31}\,\text{kg}} \left(\frac{1^2}{(200\,\text{pm})^2} + \frac{2^2}{(200\,\text{pm})^2} + \frac{1^2}{(200\,\text{pm})^2}\right)$
$= \frac{(6.63 \times 10^{-34}\,\text{J}\,\text{s})^2}{8 \times 9.11 \times 10^{-31}\,\text{kg} \times (200 \times 10^{-12}\,\text{m})^2} \left(1^2 + 2^2 + 1^2\right)$
$= \frac{43.9569 \times 10^{-68}\,\text{J}^2\,\text{s}^2}{72.88 \times 10^{-31}\,\text{kg} \times 4 \times 10^4 \times 10^{-24}\,\text{m}^2} \times (1 + 4 + 1) = \frac{0.1508 \times 10^{-17}\,(\text{kg}\,\text{m}^2\,\text{s}^{-2})^2\,\text{s}^2}{\text{kg}\,\text{m}^2} \times 6$
$= 9.048 \times 10^{-18}\,\frac{\text{kg}^2\,\text{m}^4\,\text{s}^{-4}\,\text{s}^2}{\text{kg}\,\text{m}^2} = 9.048 \times 10^{-18}\,\text{kg}\,\text{m}^2\,\text{s}^{-2} = 9.05 \times 10^{-18}\,\text{J} = 9.05\,\text{aJ}$

Chapter 19

1. a. $9 = 3^2, 2 = \log_3 9$
b. $16 = 4^2, 2 = \log_4 16$

c. $16 = 2^4$, $4 = \log_2 16$
d. $27 = 3^3$, $3 = \log_3 27$
e. $125 = 5^3$, $3 = \log_5 125$

2. a. $\ln 2.5 = 0.916$
b. $\ln 6.37 = 1.852$
c. $\ln 1.0 = 0.0$
d. $\ln 0.256 = -1.363$
e. $\ln 0.001 = -6.907$

3. a. $\ln 20 = \ln 4 + \ln 5 = 1.386 + 1.609 = 2.995 = \ln 20$
b. $\ln 10 = \ln 2 + \ln 5 = 0.693 + 1.609 = 2.302 = \ln 10$
c. $\ln 10 = \ln 30 - \ln 3 = 3.401 - 1.099 = 2.303 = \ln 10$
d. $\ln 6 = \ln 18 - \ln 3 = 2.890 - 1.099 = 1.791 = \ln 6$
e. $\ln 9 = 2 \ln 3 = 2 \times 1.099 = 2.197 = \ln 9$

4. $S = k \ln W = 1.381 \times 10^{-23}\ \mathrm{J\ K^{-1}} \times \ln 6 = 1.381 \times 10^{-23}\ \mathrm{J\ K^{-1}} \times 1.792 = 2.474 \times 10^{-23}\ \mathrm{J\ K^{-1}}$

5. $\Delta G^{\ominus} = -8.314\ \mathrm{J\ K^{-1}\ mol^{-1}} \times 298\ \mathrm{K} \times \ln(1.8 \times 10^{-5}) = -8.314\ \mathrm{J\ K^{-1}\ mol^{-1}} \times 298\ \mathrm{K} \times (-10.925) = 27\,067\ \mathrm{J\ mol^{-1}} = 27.1\ \mathrm{kJ\ mol^{-1}}$

Chapter 20

1. a. $\log 10 = 1$
b. $\log 10^4 = 4$
c. $\log 10^8 = 8$
d. $\log 10^{-3} = -3$
e. $\log 10^{-6} = -6$

2. a. $\ln 10^2 = 2.303 \log 10^2 = 2.303 \times 2 = 4.606$
b. $\ln 10^5 = 2.303 \log 10^5 = 2.303 \times 5 = 11.52$
c. $\ln 10^{10} = 2.303 \log 10^{10} = 2.303 \times 10 = 23.03$
d. $\ln 10^{-7} = 2.303 \log 10^{-7} = 2.303 \times (-7) = -16.121$
e. $\ln 0.01 = \ln 10^{-2} = 2.303 \log 10^{-2} = 2.303 \times (-2) = -4.606$

3. a. $\log 4.18 = 0.621$
b. $\log (3.16 \times 10^4) = 4.50$
c. $\log (7.91 \times 10^{-4}) = -3.102$
d. $\log 0.003\,27 = -2.485$
e. $\log 3028 = 3.481$

4. a. $\mathrm{pH} = -\log 0.01 = -\log 10^{-2} = -(-2.0) = 2.0$
b. $\mathrm{pH} = -\log 0.002 = -(-2.699) = 2.699$
c. $\mathrm{pH} = -\log 5.0 = -0.699$
d. $\mathrm{pH} = -\log 0.1014 = -(-0.994) = 0.994$
e. $\mathrm{pH} = -\log 1.072 = -0.030$

5. $\log \gamma_\pm = -(0.509\ \mathrm{kg^{1/2}\ mol^{-1/2}})|z_+z_-|\sqrt{I}$
$\ln \gamma_\pm = 2.303 \times (-0.509\ \mathrm{kg^{1/2}\ mol^{-1/2}})|z_+z_-|\sqrt{I}$
$= -(1.172\ \mathrm{kg^{1/2}\ mol^{-1/2}})|z_+z_-|\sqrt{I}$

6. $T = \dfrac{100 - 60}{60} = \dfrac{40}{100} = 0.40$
$A = -\log T = -\log 0.4 = -(-0.398) = 0.398$

Chapter 21

1. a. $e^2 = 7.39$
b. $e^{10} = 2.2 \times 10^4$
c. $e^{1.73} = 5.64$

d. $e^{2.65} = 14.2$
e. $e^{9.9} = 19\,930$

2. a. $e^{-3} = 0.050$
b. $e^{-7} = 9.12 \times 10^{-4}$
c. $e^{-2.19} = 0.112$
d. $e^{-3.83} = 0.0217$
e. $e^{-4.7} = 9.10 \times 10^{-3}$

3. $\Psi = \left(\frac{1}{3.142}\right)^{1/2} \left(\frac{1}{5.292 \times 10^{-11}\,\text{m}}\right) e^{-2.42/5.292}$
$= 0.318^{1/2} \times (1.890 \times 10^{10}\,\text{m}^{-1})^{3/2} \times e^{-0.459}$
$= 0.564 \times 2.60 \times 10^{15} \times 0.632 = 9.26 \times 10^{14}\,\text{m}^{-3/2}$

4. $n = n_0/2$, so $n = n_0 e^{-1.54\times10^{-10}\text{yr}^{-1}\times4.51\times10^{9}\text{yr}} = n_0 e^{-0.695} = 0.50 n_0$
$n/n_0 = 0.50$

Chapter 22

1. $arc\, f(x) = (x - 7)/4$

2. $arc\, f(x) = (e^x - 1)/3$

3. $arc\, g(y) = \frac{1}{3} \ln\left(\frac{y}{2}\right)$

4. $\ln c - \ln c_0 = -kt$, $\ln c_0 - \ln c = kt$, $\ln c_0/c = kt$, $c_0/c = e^{kt}$ or $c = c_0 e^{-kt}$

5. $E - E^{\ominus} = -\frac{RT}{nF} \ln Q$
$\ln Q = -\frac{nF}{RT}(E - E^{\ominus}) = -\frac{2 \times 96\,485\,\text{C mol}^{-1}}{8.314\,\text{J K}^{-1}\,\text{mol}^{-1} \times 298\,\text{K}}(-0.029\,\text{V} - 0.021\,\text{V})$
$= 77.89\,\text{C J}^{-1} \times (0.050\,\text{V})$
$= -77.89\,\text{V}^{-1} \times 0.050\,\text{V} = -3.895$
$Q = e^{3.895} = 49.1$

6. $\log \gamma_{\pm} = -0.51 \times |2 \times 1| \times \sqrt{0.125} = -0.51 \times 2 \times 0.354 = -0.361$
$\gamma_{\pm} = 10^{-0.361} = 0.436$

Chapter 23

1. a. $y = 5x + 2$, $m = 5$, $c = 2$
b. $y = 3x - 7$, $m = 3$, $c = -7$
c. $2y = 4x - 9$, $y = (4x - 9)/2 = 2x - 9/2$, $m = 2$, $c = -9/2$
d. $x + y = 2$, $y = -x + 2$, $m = -1$, $c = 2$
e. $2x + 3y = 8$, $3y = -2x + 8$, $y = -2x/3 + 8/3$, $m = -2/3$, $c = 8/3$

2. a. y against x^2, $m = 3$, $c = -8$
b. y^2 against x, $m = 5$, $c = -4$
c. y against $1/x$, $m = 2$, $c = -3$
d. y^2 against $1/x$, $m = 3$, $c = 6$
e. y^2 against $1/x^2$, $m = 2$, $c = 0$

3. a. $x^2 + y^2 = 9$, $y^2 = -x^2 + 9$, y^2 against x^2, $m = -1$, $c = 9$
b. $2x^2 - y^2 = 5$, $y^2 = 2x^2 - 5$, y^2 against x^2, $m = 2$, $c = -5$
c. $xy = 10$, $y = 10/x$, y against $1/x$, $m = 10$, $c = 0$
d. $x^2 y = 4$, $y = 4/x^2$, y against $1/x^2$, $m = 4$, $c = 0$
e. $xy^2 - 14 = 0$, $xy^2 = 14$, $y^2 = 14/x$, y^2 against $1/x$, $m = 14$, $c = 0$

4. a. $c_0 - c = kt$, $c = -kt + c_0$, c against t, $m = -k$, intercept $= c_0$
b. $1/c = kt + 1/c_0$, $1/c$ against t, $m = k$, intercept $= 1/c_0$

5. Plot $\ln p/p^\ominus$ against $1/T$, $m = -\Delta_{vap}H/R$

6. Plot $\ln k$ against $1/T$, $m = -E_a/R$

Chapter 24

1. a. $x^2 + 3x - 10 = 0$, $(x + 5)(x - 2) = 0$, $x = 2$ or $x = -5$
   b. $x^2 - 3x = 0$, $x(3 - x) = 0$, $x = 0$ or $x = 3$
   c. $3x^2 - 2x - 1 = 0$, $(3x + 1)(x - 1) = 0$, $x = 1$ or $3x = -1$, $x = -1/3$

2. a. $2x^2 - 9x + 2 = 0$, $a = 2$, $b = -9$, $c = 2$

$$x = \frac{9 \pm \surd(81 - 4 \times 2 \times 2)}{2 \times 2} = \frac{9 \pm \surd(81 - 16)}{4} = \frac{9 \pm \surd(65)}{4} = \frac{9 \pm 8.062}{4}$$
$$= \frac{17.062}{4} = 4.266 \quad \text{or} \quad \frac{0.938}{4} = 0.235$$

   b. $4x^2 + 4x + 1 = 0$, $a = 4$, $b = 4$, $c = 1$

$$x = \frac{-4 \pm \surd(16 - 4 \times 4 \times 1)}{2 \times 4} = \frac{-4 \pm \surd(16 - 16)}{8} = \frac{-4}{8} = -1/2$$

   Note that this equation has two identical solutions
   c. $3.6x^2 + 1.2x - 0.8 = 0$, $a = 3.6$, $b = 1.2$, $c = -0.8$

$$x = \frac{-1.2 \pm \surd[1.2^2 - 4 \times 3.6 \times (-0.8)]}{2 \times 3.6} = \frac{-1.2 \pm \surd(1.44 + 11.52)}{7.2} = \frac{-1.2 \pm \surd 12.96}{7.2} = \frac{-1.2 \pm 3.6}{7.2}$$
$$= \frac{2.4}{7.2} = 0.333 \quad \text{or} \quad -\frac{4.8}{7.2} = -0.667$$

3. $a = 0.04$, $b = 1.715 \times 10^{-3}$, $c = -1.715 \times 10^{-3}$

$$\alpha = \frac{-1.715 \times 10^{-3} \pm \surd[(1.715 \times 10^{-3})^2 - 4 \times 0.04 \times (-1.715 \times 10^{-3})]}{2 \times 0.04}$$
$$= \frac{-1.715 \times 10^{-3} \pm \surd(2.942 \times 10^{-6} + 2.744 \times 10^{-4})}{0.08} = \frac{-1.715 \times 10^{-3} \pm \surd(2.773 \times 10^{-4})}{0.08}$$
$$= \frac{-1.715 \times 10^{-3} \pm 1.665 \times 10^{-2}}{0.08}$$
$$= \frac{1.494 \times 10^{-2}}{0.08} = 0.187 \quad \text{or} \quad \frac{-1.837 \times 10^{-2}}{0.08} = -0.230$$

   Since α must be positive, the solution is $\alpha = 0.187$

4. $a = 1$, $b = 1.75 \times 10^{-5}$, $c = -1.75 \times 10^{-6}$

$$m = \frac{-1.75 \times 10^{-5} \pm \surd[(1.75 \times 10^{-5})^2 - 4 \times 1 \times (-1.75 \times 10^{-6})]}{2 \times 1}$$
$$= \frac{-1.75 \times 10^{-5} \pm \surd(3.063 \times 10^{-10} + 7 \times 10^{-6})}{2} = \frac{-1.75 \times 10^{-5} \pm \surd(7.00 \times 10^{-6})}{2}$$
$$= \frac{-1.75 \times 10^{-5} \pm 2.646 \times 10^{-3}}{2}$$
$$= \frac{2.629 \times 10^{-3}}{2} = 1.315 \times 10^{-3} \quad \text{or} \quad \frac{-2.664 \times 10^{-3}}{2} = -1.332 \times 10^{-3}$$

   Since m must be positive, the solution is $m = 1.315 \times 10^{-3}$

5. $a = 4.0540, b = 2.1750, c = 0.1054$

$$\alpha = \frac{-2.1750 \pm \sqrt{(2.1750^2 - 4 \times 4.0540 \times 0.1054)}}{2 \times 4.0540} = \frac{-2.1750 \pm \sqrt{(4.7306 - 1.7092)}}{8.108}$$
$$= \frac{-2.1750 \pm \sqrt{3.0214}}{8.108} = \frac{-2.1750 \pm 1.7382}{8.108}$$
$$= \frac{-0.4368}{8.108} = -0.0539 \quad \text{or} \quad \frac{-3.9132}{8.108} = -0.483$$

Since the system has been perturbed, a negative value of α denotes a shift of the equilibrium to the left. However, in this case there is no obvious way of telling which solution is correct.

Chapter 25

1. a. For $x = 0.5$, $1 + x + 2x^2 + 3x^3 + 4x^4 + 5x^5$
$= 1 + 0.5 + 2 \times 0.5^2 + 3 \times 0.5^3 + 4 \times 0.5^4 + 5 \times 0.5^5$
$= 1 + 0.5 + 2 \times 0.25 + 3 \times 0.125 + 4 \times 0.0625 + 5 \times 0.031\,25$
$= 1 + 0.5 + 0.5 + 0.375 + 0.25 + 0.156\,25$
i.e. converges as terms get smaller

b. For $x = 2$, $1 + \frac{1}{x!} + \frac{2}{(2x)!} + \frac{3}{(3x)!} + \frac{4}{(4x)!}$
$= 1 + \frac{1}{2!} + \frac{2}{(2 \times 2)!} + \frac{3}{(3 \times 2)!} + \frac{4}{(4 \times 2)!}$
$= 1 + \frac{1}{2} + \frac{2}{4!} + \frac{3}{6!} + \frac{4}{8!} = 1 + \frac{1}{2} + \frac{2}{24} + \frac{3}{720} + \frac{4}{40\,320}$
$= 1 + 0.5 + 0.083 + 0.0042 + 0.0001$
i.e. converges as terms get smaller

c. For $x = 0.5$, $1 + \ln x + 2 \ln 2x + 3 \ln 3x + 4 \ln 4x = 1 + \ln 0.5 + 2 \ln 1 + 3 \ln 1.5 + 4 \ln 2$
$= 1 + (-0.693) + 2 \times 0 + 3 \times 0.405 + 4 \times 0.693 = 1 + (-0.693) + 0 + 1.215 + 2.772$
i.e. diverges as terms get larger

2. a. $1 + e^{-4} + e^{-2\times 4} + e^{-3\times 4} = 1 + e^{-4} + e^{-8} + e^{-12} = 1 + 0.0183 + 3.35 \times 10^{-4} + 6.14 \times 10^{-6}$
Sums are 1, 1.0183, 1.0186, 1.0186; limit is 1.019

b. $\frac{1}{4} + \frac{1}{2 \times 4^2} + \frac{1}{3 \times 4^3} + \frac{1}{4 \times 4^4} + \frac{1}{5 \times 4^5}$
$= \frac{1}{4} + \frac{1}{2 \times 16} + \frac{1}{3 \times 64} + \frac{1}{4 \times 256} + \frac{1}{5 \times 1024}$
$= \frac{1}{4} + \frac{1}{32} + \frac{1}{192} + \frac{1}{1024} + \frac{1}{5120}$
$= 0.25 + 0.031 + 5.21 \times 10^{-3} + 9.77 \times 10^{-4} + 1.95 \times 10^{-4}$
Sums are 0.281, 0.286, 0.287, 0.287; limit is 0.287

c. $\frac{1}{\ln 4} + \frac{1}{(\ln 8)^2} + \frac{1}{(\ln 12)^3} + \frac{1}{(\ln 16)^4} + \frac{1}{(\ln 20)^5} + \frac{1}{(\ln 24)^6}$
$= \frac{1}{1.386} + \frac{1}{2.079^2} + \frac{1}{2.484^3} + \frac{1}{2.772^4} + \frac{1}{2.996^5} + \frac{1}{3.178^6}$
$= \frac{1}{1.386} + \frac{1}{4.324} + \frac{1}{15.327} + \frac{1}{59.04} + \frac{1}{243} + \frac{1}{1030}$
$= 0.722 + 0.231 + 0.065 + 0.017 + 0.004 + 9.71 \times 10^{-4}$
Sums are 0.953, 1.018, 1.035, 1.039, 1.040; limit is 1.044

3. $\frac{B}{V} = \frac{-4.5 \times 10^{-6}\,\text{m}^3\,\text{mol}^{-1}}{0.025\,\text{m}^3\,\text{mol}^{-1}} = -1.8 \times 10^{-4}$
$\frac{C}{V^2} = \frac{1.10 \times 10^{-9}\,\text{m}^6\,\text{mol}^{-1}}{(0.025\,\text{m}^3\,\text{mol}^{-1})^2} = \frac{1.10 \times 10^{-9}\,\text{m}^6\,\text{mol}^{-2}}{6.25 \times 10^{-4}\,\text{m}^6\,\text{mol}^{-2}} = 1.76 \times 10^{-6}$
$\frac{B/V}{C/V^2} = \frac{-1.8 \times 10^{-4}}{1.76 \times 10^{-6}} = -102$

4. $\dfrac{\varepsilon_1}{kT} = \dfrac{3.15 \times 10^{-21}\,\mathrm{J}}{1.38 \times 10^{-23}\,\mathrm{J\,K^{-1}} \times 298\,\mathrm{K}} = 0.766$

$\dfrac{\varepsilon_2}{kT} = \dfrac{4.5 \times 10^{-23}\,\mathrm{J}}{1.38 \times 10^{-23}\,\mathrm{J\,K^{-1}} \times 298\,\mathrm{K}} = 0.0109$

$$q_e = 5e^{-0/kT} + 3e_1^{-\varepsilon_1/kT} + e_2^{-\varepsilon_2/kT} = 5e^0 + 3e^{-0.766} + e^{-0.0109}$$
$$= 5 \times 1 + 3 \times 0.465 + 0.989 = 5 + 1.395 + 0.989 = 7.38$$

5. $q_r = (2 \times 0 + 1)e^{-0(0+1)h^2/8\pi^2 IkT} + (2 \times 1 + 1)e^{-1(1+1)h^2/8\pi^2 IkT} + (2 \times 2 + 1)e^{-2(2+1)h^2/8\pi^2 IkT}$

$$+ (2 \times 3 + 1)e^{-3(3+1)h^2/8\pi^2 IkT} + (2 \times 4 + 1)e^{-4(4+1)h^2/8\pi^2 IkT}$$
$$= e^0 + 3e^{-2h^2/8\pi^2 IkT} + 5e^{-6h^2/8\pi^2 IkT} + 7e^{-12h^2/8\pi^2 IkT} + 9e^{-20h^2/8\pi^2 IkT}$$
$$= 1 + 3e^{-h^2/4\pi^2 IkT} + 5e^{-3h^2/4\pi^2 IkT} + 7e^{-3h^2/2\pi^2 IkT} + 9e^{-5h^2/2\pi^2 IkT}$$

Chapter 26

1. a. $\sin 48^\circ = 0.7431$
   b. $\cos 63^\circ = 0.4540$
   c. $\tan 57^\circ = 1.5399$
   d. $\sin(-32^\circ) = -0.5299$
   e. $\cos 171^\circ = -0.9877$

2. a. $\sin(\pi/3) = 0.8661$
   b. $\cos(3/2) = 0.0707$
   c. $\tan 2 = -2.1850$
   d. $\sin(-\pi/7) = -0.4339$
   e. $\cos(5\pi/4) = -0.7067$

3. $d = \dfrac{n\lambda}{2\sin\theta} = \dfrac{1 \times 154\,\mathrm{pm}}{2 \times \sin 12^\circ} = \dfrac{154\,\mathrm{pm}}{2 \times 0.2079} = \dfrac{154\,\mathrm{pm}}{0.4158} = 370\,\mathrm{pm}$

4. For $n = 1$, $\Psi(0) = \Psi(a) = 0$, $\Psi(a/2) = 1$ (see Fig. 26.5)
   For $n = 2$, $\Psi(a/4) = 1$, $\Psi(a/2) = 0$, $\Psi(3a/4) = -1$ (see Fig. 26.6)
   For $n = 3$, $\Psi(a/6) = \Psi(5a/6) = 1$, $\Psi(a/3) = \Psi(2a/3) = 0$, $\Psi(a/2) = -1$ (see Fig. 26.7)

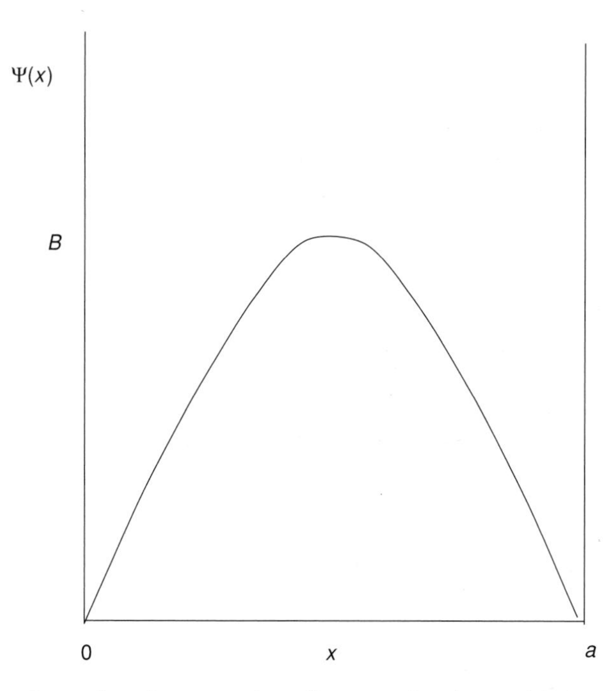

Figure 26.5 Graph of wavefunction Ψ against distance x for the particle in a box with n = 1

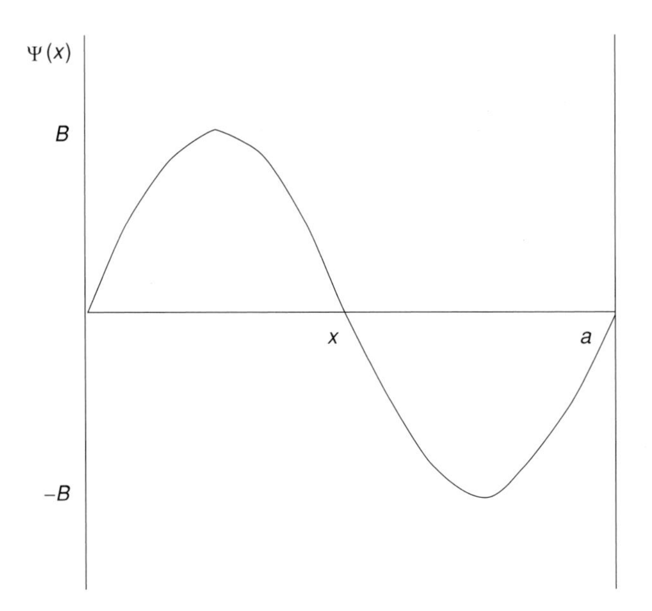

Figure 26.6 Graph of wavefunction Ψ against distance x for the particle in a box with n = 2

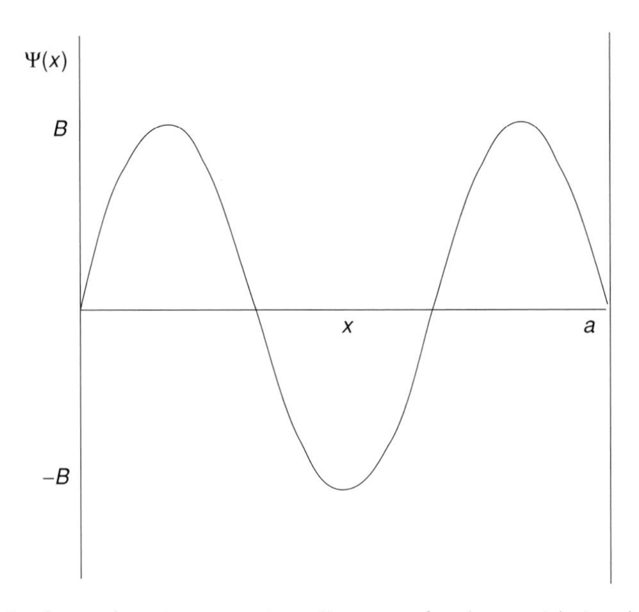

Figure 26.7 Graph of wavefunction Ψ against distance x for the particle in a box with n = 3

5. $x' = 4\cos 35^\circ - 7\sin 35^\circ = 4\times 0.8192 - 7\times 0.5736 = 3.2768 - 4.0152 = -0.7384$
   $y' = 4\sin 35^\circ + 7\cos 25^\circ = 4\times 0.5736 + 7\times 0.8192 = 2.2944 + 5.7344 = 8.0288$

6. $r^2 = 5\times 10^3 \times (154\,\text{pm})^2 \left(\dfrac{1-\cos 109.5^\circ}{1+\cos 109.5^\circ}\right) = 5\times 10^3 \times (154\times 10^{-12}\,\text{m})^2 \left(\dfrac{1-(-0.3338)}{1+(-0.3338)}\right)$

   $= 5\times 10^3 \times 2.372\times 10^{-20}\,\text{m}^2 \left(\dfrac{1+0.3338}{1-0.3338}\right) = 1.186\times 10^{-16}\,\text{m}^2 \times \dfrac{1.3338}{0.6662} = 2.374\times 10^{-16}\,\text{m}^2$

   $r = \surd(2.374\times 10^{-16}\,\text{m}^2) = 1.541\times 10^{-8}\,\text{m}$

Chapter 27

1. a. $\theta = \sin^{-1} 0.734 = 47.2^\circ$
   b. $\theta = \cos^{-1}(-0.214) = 102^\circ$
   c. $\theta = \tan^{-1} 4.78 = 78.2^\circ$
   d. $\theta = \sin^{-1}(-0.200) = -11.5^\circ$
   e. $\theta = \tan^{-1}(-2.79) = -70.3^\circ$

2. a. $x = \sin^{-1} 0.457 = 0.475$ rad
   b. $x = \cos^{-1} 0.281 = 1.29$ rad
   c. $x = \tan^{-1} 10.71 = 1.48$ rad
   d. $x = \sin^{-1}(-0.842) = -1.00$ rad
   e. $x = \cos^{-1}(-0.821) = 2.53$ rad

3. a. $\sin x = 0.104 + 0.815 = 0.919$, $x = \sin^{-1} 0.919 = 66.8^\circ = 1.17$ rad
   b. $\cos x = 0.817 - 0.421 = 0.396$, $x = \cos^{-1} 0.396 = 66.7^\circ = 1.16$ rad
   c. $3x = \tan^{-1} 5.929 = 80.43^\circ$ or 1.404 rad, $x = 80.43^\circ/3 = 26.81^\circ$ or 1.404 rad/3 = 0.468 rad
   d. $\sin 2x = 0.520 + 0.318 = 0.838$, $2x = \sin^{-1} 0.838 = 56.9^\circ$ or 0.994 rad, $x = 56.9^\circ/2 = 28.5^\circ$ or 0.994 rad/2 = 0.497 rad
   e. $\cos 4x = 0.957 - 0.212 = 0.745$, $4x = \cos^{-1} 0.745 = 41.8^\circ$ or 0.730 rad, $x = 41.8^\circ/4 = 10.5^\circ$ or 0.730 rad/4 = 0.183 rad

4. $$\sin\theta = \frac{132\text{ pm}}{2 \times 220\text{ pm}} = \frac{132\text{ pm}}{440\text{ pm}} = 0.300$$
$$\theta = \sin^{-1}\ 0.300 = 17.5^\circ$$

5. $$\sin^2\theta = \frac{(136\text{ pm})^2}{4 \times (313\text{ pm})^2}\left(1^2 + 1^2 + 2^2\right) = \frac{18\,496}{4 \times 97\,969}(1 + 1 + 4) = \frac{18\,496 \times 6}{4 \times 97\,969} = \frac{110\,976}{391\,876} = 0.283$$
$$\sin\theta = \sqrt{}0.283 = 0.532$$
$$\theta = \sin^{-1}\ 0.532 = 32.1^\circ$$

6. $$\cos\theta = -\frac{4\pi\varepsilon_0\varepsilon r^2 E}{z_A e\mu}$$
$$= -\frac{4 \times 3.142 \times 8.85 \times 10^{-12}\text{ C}^2\text{ N}^{-1}\text{ m}^{-2} \times 1.5 \times (250\text{ pm})^2 \times (-0.014 \times 10^{-18}\text{ J})}{2 \times 1.6 \times 10^{-19}\text{ C} \times 2.50 \times 10^{-30}\text{ C m}}$$
$$= -\frac{4 \times 3.142 \times 8.85 \times 10^{-12}\text{ C}^2\text{ N}^{-1}\text{ m}^{-2} \times 1.5 \times 6.25 \times 10^{-20}\text{ m}^2 \times 0.014 \times 10^{-18}\text{ N m}}{2 \times 1.6 \times 10^{-19}\text{ C} \times 2.50 \times 10^{-30}\text{ C m}}$$
$$= -\frac{1.460 \times 10^{-49}}{8.0 \times 10^{-49}} = 0.1825$$
$$\theta = \cos^{-1}(-0.1825) = 100.5^\circ$$

Chapter 28

1. a. $d = \sqrt{}[(1-4)^2 + (2-0)^2 + (3-7)^2] = \sqrt{}[(-3)^2 + 2^2 + (-4)^2] = \sqrt{}(9 + 4 + 16) = \sqrt{}29 = 5.4$
   b. $d = \sqrt{}[(2-(-4))^2 + (0-3)^2 + (4-(-2))^2] = \sqrt{}(6^2 + 3^2 + 6^2) = \sqrt{}(36 + 9 + 36) = \sqrt{}81 = 9$
   c. $d = \sqrt{}[(8-7)^2 + (-2-(-2))^2 + (-5-(-7))^2] = \sqrt{}[1^2 + 0^2 + (-2)^2] = \sqrt{}(1 + 0 + 4) = \sqrt{}5 = 2.2$

2. a. $r = \sqrt{}(1^2 + 2^2 + 3^2) = \sqrt{}(1 + 4 + 9) = \sqrt{}14 = 3.74$
   $\theta = \cos^{-1}\left(\frac{3}{3.74}\right) = \cos^{-1}\ 0.802 = 36.7^\circ$
   $\phi = \tan^{-1}\left(\frac{2}{1}\right) = \tan^{-1}\ 2 = 63.4^\circ$
   b. $r = \sqrt{}(8^2 + 7^2 + 4^2) = \sqrt{}(64 + 49 + 16) = \sqrt{}129 = 11.4$
   $\theta = \cos^{-1}\left(\frac{4}{11.4}\right) = \cos^{-1}\ 0.351 = 69.5^\circ$
   $\phi = \tan^{-1}\left(\frac{7}{8}\right) = \tan^{-1}\ 0.875 = 41.2^\circ$
   c. $r = \sqrt{}[(-1)^2 + 0^2 + (-9)^2] = \sqrt{}(1 + 81) = \sqrt{}82 = 9.1$
   $\theta = \cos^{-1}\left(\frac{-9}{9.06}\right) = \cos^{-1}(-0.993) = 173.2^\circ$
   $\phi = \tan^{-1}\left(\frac{0}{-1}\right) = \tan^{-1}\ 0 = 0.0^\circ$

3. a. $x = r\sin\theta\cos\phi = 6 \times \sin\pi/2 \times \cos\pi = 6 \times 1 \times (-1) = -6$
   $y = r\sin\theta\sin\phi = 6 \times \sin\pi/2 \times \sin\pi = 6 \times 1 \times 0 = 0$
   $z = r\cos\theta = 6\cos\pi/2 = 6 \times 0 = 0$
   b. $x = r\sin\theta\cos\phi = 10\sin(-\pi/3)\cos 2\pi = 10 \times (-0.8660) \times 1 = -8.66$
   $y = r\sin\theta\sin\phi = 10\sin(-\pi/3)\sin 2\pi = 10 \times (-0.8660) \times 0 = 0$
   $z = r\cos\theta = 10 \times \cos(-\pi/3) = 10 \times 0.5 = 5$
   c. $x = r\sin\theta\cos\phi = 7.14 \times \sin(35^\circ) \times \cos(-27^\circ) = 7.14 \times 0.5736 \times 0.8910 = 3.65$
   $y = r\sin\theta\cos\phi = 7.14 \times \sin 35^\circ \times \sin(-27^\circ) = 7.14 \times 0.5736 \times (-0.4540) = -1.86$
   $z = r\cos\theta = 7.14\cos 35^\circ = 7.14 \times 0.8192 = 5.85$

4. Interatomic distances:

	C2	C3	C4	C5	C6	C7	C8
C1	3.57	2.52	3.57	2.52	2.52	1.55	3.57
C2	—	2.52	5.05	2.52	4.37	2.96	5.05
C3	—	—	2.52	2.52	2.52	1.55	4.37
C4	—	—	—	4.37	2.52	2.96	6.18
C5	—	—	—	—	2.52	1.55	2.52
C6	—	—	—	—	—	1.55	2.52
C7	—	—	—	—	—	—	2.96

There are bonds C1–C7, C3–C7, C5–C7 and C6–C7.

5. $r = \frac{a_0}{2} = \frac{5.292 \times 10^{-11}\ \text{m}}{2} = 2.646 \times 10^{-11}\ \text{m}$
$x = r\sin\theta\cos\phi = 2.646 \times 10^{-11} \times \sin 45° \times \cos 45° = 2.646 \times 10^{-11} \times 0.7071 \times 0.7071 = 1.323 \times 10^{-11}$
$y = r\sin\theta\sin\phi = 2.646 \times 10^{-11} \times \sin 45° \times \sin 45° = 2.646 \times 10^{-11} \times 0.7071 \times 0.7071 = 1.323 \times 10^{-11}$
$z = r\cos\theta = 2.646 \times 10^{-11} \times \cos 45° = 2.646 \times 10^{-11} \times 0.7071 = 1.871 \times 10^{-11}$

6. $r = a_0 = 5.292 \times 10^{-11}\ \text{m}$
$x = r\sin\theta\cos\phi = 5.292 \times 10^{-11}\ \text{m} \times \sin 60° \times \cos(-30°)$
$= 5.292 \times 10^{-11} \times 0.8660 \times 0.8660 = 3.969 \times 10^{-11}$
$y = r\sin\theta\sin\phi = 5.292 \times 10^{-11} \times \sin 60° \times \sin(-30°)$
$= 5.292 \times 10^{-11} \times 0.8660 \times (-0.50) = -2.292 \times 10^{-11}$
$z = r\cos\theta = 5.292 \times 10^{-11} \times \cos 60° = 5.292 \times 10^{-11} \times 0.50 = 2.646 \times 10^{-11}$

Chapter 29

1. a. $|3\mathbf{i} + 8\mathbf{j} + 9\mathbf{k}| = \sqrt{(3^2 + 8^2 + 9^2)} = \sqrt{(9 + 64 + 81)} = \sqrt{154} = 12.4$
   b. $|5\mathbf{i} - 8\mathbf{j} + 8\mathbf{k}| = \sqrt{(5^2 + 8^2 + 8^2)} = \sqrt{(25 + 64 + 64)} = \sqrt{153} = 12.4$
   c. $|-2\mathbf{i} + 9\mathbf{j} - 4\mathbf{k}| = \sqrt{(2^2 + 9^2 + 4^2)} = \sqrt{(4 + 81 + 16)} = \sqrt{101} = 10.0$

2. a. $(5 + 3)\mathbf{i} + (6 + 6)\mathbf{j} + (9 + 2)\mathbf{k} = 8\mathbf{i} + 12\mathbf{j} + 11\mathbf{k}$
   b. $(2 + 5)\mathbf{i} + (-6 + 0)\mathbf{j} + (9 - 8)\mathbf{k} = 7\mathbf{i} - 6\mathbf{j} + \mathbf{k}$
   c. $(9 + 5)\mathbf{i} + (2 - 3)\mathbf{j} + (-2 + 8)\mathbf{k} = 14\mathbf{i} - \mathbf{j} + 6\mathbf{k}$

3. a. $(5 - 3)\mathbf{i} + (6 - 6)\mathbf{j} + (9 - 2)\mathbf{k} = 2\mathbf{i} + 7\mathbf{k}$
   b. $(2 - 5)\mathbf{i} + (-6 - 0)\mathbf{j} + (9 - (-8))\mathbf{k} = -3\mathbf{i} - 6\mathbf{j} + 17\mathbf{k}$
   c. $(9 - 5)\mathbf{i} + (2 - (-3))\mathbf{j} + (-2 - 8)\mathbf{k} = 4\mathbf{i} + 5\mathbf{j} - 10\mathbf{k}$

4. a. $|2\mathbf{i} + 6\mathbf{j} - 9\mathbf{k}| = \sqrt{(2^2 + 6^2 + 9^2)} = \sqrt{(4 + 36 + 81)} = \sqrt{121} = 11$
   $\hat{\mathbf{n}} = \frac{1}{11}(2\mathbf{i} + 6\mathbf{j} - 9\mathbf{k}) = 0.18\mathbf{i} + 0.54\mathbf{j} - 0.81\mathbf{k}$
   b. $|5\mathbf{i} - 8\mathbf{j} + 7\mathbf{k}| = \sqrt{(5^2 + 8^2 + 7^2)} = \sqrt{(25 + 64 + 49)} = \sqrt{138} = 11.7$
   $\hat{\mathbf{n}} = \frac{1}{11.7}(5\mathbf{i} - 8\mathbf{j} + 7\mathbf{k}) = 0.43\mathbf{i} - 0.68\mathbf{j} + 0.60\mathbf{k}$
   c. $|2\mathbf{i} + \mathbf{j} - 6\mathbf{k}| = \sqrt{(2^2 + 1^2 + 6^2)} = \sqrt{(4 + 1 + 36)} = \sqrt{41} = 6.4$
   $\hat{\mathbf{n}} = \frac{1}{6.4}(2\mathbf{i} + \mathbf{j} - 6\mathbf{k}) = 0.31\mathbf{i} + 0.16\mathbf{j} - 0.94\mathbf{k}$

5. $\mathbf{r}_2 - \mathbf{r}_1 = (x_2 - x_1)\mathbf{i} + (y_2 - y_1)\mathbf{j} + (z_2 - z_1)\mathbf{k}$

   $|\mathbf{r}_2 - \mathbf{r}_1| = \sqrt{[(x_2 - x_1)^2 + (y_2 - y_1)^2 + (z_2 - z_1)^2]}$

6. $\mathbf{a} = |\mathbf{a}|\mathbf{i}$, $\mathbf{b} = |\mathbf{b}|\mathbf{j}$, $\mathbf{c} = |\mathbf{c}|\mathbf{k}$; $\mathbf{T} = n_1|\mathbf{a}|\mathbf{i} + n_2|\mathbf{b}|\mathbf{j} + n_3|\mathbf{c}|\mathbf{k}$

Chapter 30

1. a. $2\mathbf{a} = 2(3\mathbf{i} + \mathbf{j} - 2\mathbf{k}) = 6\mathbf{i} + 2\mathbf{j} - 4\mathbf{k}$
   b. $3\mathbf{b} = 3(5\mathbf{i} - 2\mathbf{j} + 3\mathbf{k}) = 15\mathbf{i} - 6\mathbf{j} + 9\mathbf{k}$
   c. $4.5\mathbf{c} = 4.5(0.7\mathbf{i} + 3.4\mathbf{j} + 2.1\mathbf{k}) = 3.2\mathbf{i} + 15\mathbf{j} + 9.5\mathbf{k}$

2. a. $\mathbf{a} \cdot \mathbf{b} = 3\mathbf{i} \cdot (-\mathbf{i}) + 2\mathbf{j} \cdot (-2\mathbf{j}) + 4\mathbf{k} \cdot 3\mathbf{k} = -3 - 4 + 12 = 5$
   b. $\mathbf{a} \cdot \mathbf{b} = 3\mathbf{i} \cdot (-8\mathbf{i}) + (-4\mathbf{j}) \cdot 6\mathbf{j} + (-5\mathbf{k}) \cdot 3\mathbf{k} = -24 - 24 - 15 = -63$
   c. $\mathbf{a} \cdot \mathbf{b} = 0\mathbf{i} \cdot 6\mathbf{i} + 8\mathbf{j} \cdot 4\mathbf{j} + (-7\mathbf{k}) \cdot (-5\mathbf{k}) = 0 + 32 + 35 = 67$

3. a. $\mathbf{a} \times \mathbf{b} = 3\mathbf{i} \times (-2\mathbf{j}) + 3\mathbf{i} \times 3\mathbf{k} + 2\mathbf{j} \times (-\mathbf{i}) + 2\mathbf{j} \times 3\mathbf{k} + 4\mathbf{k} \times (-\mathbf{i}) + 4\mathbf{k} \times (-2\mathbf{j})$
$= -6(\mathbf{i} \times \mathbf{j}) + 9(\mathbf{i} \times \mathbf{k}) - 2(\mathbf{j} \times \mathbf{i}) + 6(\mathbf{j} \times \mathbf{k}) - 4(\mathbf{k} \times \mathbf{i}) - 8(\mathbf{k} \times \mathbf{j})$
$= -6\mathbf{k} + 9(-\mathbf{j}) - 2(-\mathbf{k}) + 6\mathbf{i} - 4\mathbf{j} - 8(-\mathbf{i}) = 14\mathbf{i} - 13\mathbf{j} - 4\mathbf{k}$
   b. $\mathbf{a} \times \mathbf{b} = 3\mathbf{i} \times 6\mathbf{j} + 3\mathbf{i} \times 3\mathbf{k} - 4\mathbf{j} \times (-8\mathbf{i}) - 4\mathbf{j} \times 3\mathbf{k} - 5\mathbf{k} \times (-8\mathbf{i}) - 5\mathbf{k} \times 6\mathbf{j}$
$= 18(\mathbf{i} \times \mathbf{j}) + 9(\mathbf{i} \times \mathbf{k}) + 32(\mathbf{j} \times \mathbf{i}) - 12(\mathbf{j} \times \mathbf{k}) + 40(\mathbf{k} \times \mathbf{i}) - 30(\mathbf{k} \times \mathbf{j})$
$= 18\mathbf{k} - 9\mathbf{j} - 32\mathbf{k} - 12\mathbf{i} + 40\mathbf{j} + 30\mathbf{i} = 18\mathbf{i} + 31\mathbf{j} - 14\mathbf{k}$
   c. $\mathbf{a} \times \mathbf{b} = 8\mathbf{j} \times 6\mathbf{i} + 8\mathbf{j} \times (-5\mathbf{k}) - 7\mathbf{k} \times 6\mathbf{i} - 7\mathbf{k} \times 4\mathbf{j}$
$= 48(\mathbf{j} \times \mathbf{i}) - 40(\mathbf{j} \times \mathbf{k}) - 42(\mathbf{k} \times \mathbf{i}) - 28(\mathbf{k} \times \mathbf{j})$
$= -48\mathbf{k} - 40\mathbf{i} - 42\mathbf{j} + 28\mathbf{i} = -12\mathbf{i} - 42\mathbf{j} - 48\mathbf{k}$

4. $\cos\theta = \dfrac{\mathbf{a} \cdot \mathbf{b}}{|\mathbf{a}|\,|\mathbf{b}|} = \dfrac{3.62}{2.14 \times 5.19} = \dfrac{3.62}{11.11} = 0.326, \theta = \cos^{-1} 0.326 = 71.0^\circ$

5. $E = -\boldsymbol{\mu} \cdot \boldsymbol{B} = -|\boldsymbol{\mu}||\boldsymbol{B}| \cos\theta = -9.274 \times 10^{-24}\ \text{J T}^{-1} \times 2.0\ \text{T} \times \cos 30^\circ$
$= -9.274 \times 10^{-24} \times 2.0 \times 0.8660\ \text{J} = -1.6 \times 10^{-23}\ \text{J}$

6. $\mathbf{L} = m(\mathbf{r} \times \mathbf{v}) = 9.109 \times 10^{-31}\ \text{kg}\ (5.292 \times 10^{-11}\ m\ \mathbf{i} \times 6 \times 10^6\ \text{m s}^{-1}\mathbf{j})$
$= 9.109 \times 10^{-31}\ \text{kg} \times 5.292 \times 10^{-11}\ \text{m} \times 6 \times 10^6\ \text{m s}^{-1}(\mathbf{i} \times \mathbf{j})$
$= (2.892 \times 10^{-34}\ \text{kg m}^2\ \text{s}^{-1})\mathbf{k}$

Chapter 31

1. a. $a = 2, b = 3$
   b. $a = 3, b = -6$
   c. $a = 4, b = 7$
   d. $a = 5, b = -9$
   e. $a = x, b = y$

2. a. $2 - 3i$
   b. $3 + 6i$
   c. $4 - 7i$
   d. $5 + 9i$
   e. $x - iy$

3. $F(hkl) = \sum_j f_j e^{i2\pi(hx_j + ky_j + lz_j)} = f_1 e^{i2\pi(hx_1+ky_1+lz_1)} + f_2 e^{i2\pi(hx_2+ky_2+lz_2)} + f_3 e^{i2\pi(hx_3+ky_3+lz_3)}$
$= f_1[\cos 2\pi(hx_1 + ky_1 + lz_1) + i \sin 2\pi(hx_1 + ky_1 + lz_1)] + f_2[\cos 2\pi(hx_2 + ky_2 + lz_2)$
$+ i \sin 2\pi(hx_2 + ky_2 + lz_2)] + f_3[\cos 2\pi(hx_3 + ky_3 + lz_3) + i \sin 2\pi(hx_3 + ky_3 + lz_3)]$
$= [f_1 \cos 2\pi(hx_1 + ky_1 + lz_1) + f_2 \cos 2\pi(hx_2 + ky_2 + lz_2) + f_3 \cos 2\pi(hx_3 + ky_3 + lz_3)]$
$+ i[f_1 \sin 2\pi(hx_1 + ky_1 + lz_1) + f_2 \sin 2\pi(hx_2 + ky_2 + lz_2) + f_3 \sin 2\pi(hx_3 + ky_3 + lz_3)]$

4. $\Psi = R + iI, \Psi^* = R - iI, \Psi^*\Psi = (R + iI)(R - iI) = R^2 - RiI + Ri - i^2 I^2 = R^2 - (-1)I^2$
$= R^2 + I^2$, i.e. no imaginary part because i does not appear.

5. $\Psi_{2p_x} = Ae^{-r/2a_0} r \sin\theta(\cos\phi + i \sin\phi), \Psi_{2p_y} = Ae^{-r/2a_0} r \sin\theta(\cos\phi - i \sin\phi)$
$\Psi_{2p_x} + \Psi_{2p_y} = Ae^{-r/2a_0} r \sin\theta(\cos\phi + \cos\phi + i \sin\phi - i \sin\phi) = 2Ae^{-r/2a_0} r \sin\theta \cos\phi$

Chapter 32

1. a. $m = 6 \times (-4) = -24$
   b. $m = 6 \times 0 = 0$
   c. $m = 6 \times 2 = 12$
   d. $m = 6 \times (-0.5) = -3.0$
   e. $m = 6 \times 2.42 = 14.5$

2. a. $m = 21 \times (-5)^2 - 6 \times (-5) = 21 \times 25 + 30 = 525 + 30 = 555$
   b. $m = 21 \times 0^2 - 6 \times 0 = 0$
   c. $m = 21 \times 7^2 - 6 \times 7 = 21 \times 49 - 42 = 1029 - 42 = 987$
   d. $m = 21 \times (-3.6)^2 - 6 \times (-3.6) = 21 \times 12.96 + 21.6 = 272.2 + 21.6 = 294$
   e. $m = 21 \times 7.41^2 - 6 \times 7.41 = 21 \times 54.91 - 44.46 = 1153.11 - 44.46 = 1108$

3. $m = \dfrac{d(\ln K)}{dT} = -\dfrac{\Delta H^{\circ}}{RT^2} = -\dfrac{(-119.7 \times 10^3\ \mathrm{J\ mol^{-1}})}{8.314\ \mathrm{J\ K^{-1}\ mol^{-1}} \times (298\ \mathrm{K})^2} = 0.162\ \mathrm{K^{-1}}$

4. $m = \dfrac{dV}{dt} = \dfrac{25 \times 10^3\ \mathrm{Pa} \times 3.142 \times (5 \times 10^{-3}\ \mathrm{m})^4}{8 \times 9.33 \times 10^{-6}\ \mathrm{kg\ m^{-1}\ s^{-1}} \times 10 \times 10^{-2}\ \mathrm{m}}$
$= \dfrac{4.909 \times 10^{-5}\ \mathrm{m^3\ s^{-2}}}{7.464 \times 10^{-6}\ \mathrm{s^{-1}}} = 6.57\ \mathrm{m^3\ s^{-1}}$

5. $m = \frac{dc}{dt} = -kc^2 = -0.775\ \mathrm{dm^3\ mol^{-1}\ s^{-1}} \times (0.05\ \mathrm{mol\ dm^{-3}})^2$
$= -0.775\ \mathrm{dm^3\ mol^{-1}\ s^{-1}} \times 2.5 \times 10^{-3}\ \mathrm{mol^2\ dm^{-6}} = -1.94 \times 10^{-3}\ \mathrm{mol\ dm^{-3}\ s^{-1}}$

Chapter 33

1. $\dfrac{df(x)}{dx} = 3 \times 5^{5-1} - 4 \times 3x^{3-1} + 2 \times 1 \times x^{1-0} = 3 \times 5x^4 - 4 \times 3x^2 + 2 \times x^0 = 15x^4 - 12x^2 + 2$

2. $\dfrac{dg(y)}{dy} = 2 \times 4y^{4-1} + 3 \times 3y^{3-1} - 5 \times 2y^{2-1} - 1 \times y^{1-0} = 8y^3 + 9y^2 - 10y - 1$

3. $\dfrac{dH_4(\xi)}{d\xi} = 0 - 48 \times 2\xi^{2-1} + 16 \times \xi^{4-1} = -96\xi + 64\xi^3$

4. $\dfrac{dL_3(\rho)}{d\rho} = 0 - 18 \times 1 \times \rho^{1-0} + 9 \times 2\rho^{2-1} - 3\rho^{3-1} = -18 + 18\rho - 3\rho^2$

5. $\dfrac{dP_3(z)}{dz} = \dfrac{5}{2} \times 3z^{3-1} - \dfrac{3}{2} \times 1 \times z^{1-0} = \dfrac{15}{2}z^2 - \dfrac{3}{2}$

6. $\dfrac{dV}{dT} = V_0(0 + 4.2 \times 10^{-4} \times 1 \times T^{1-0} + 1.67 \times 10^{-6} \times 2T^{2-1}) = V_0(4.2 \times 10^{-4} + 3.34 \times 10^{-6}T)$
At 350 K, $\dfrac{dV}{dT} = V_0(4.2 \times 10^{-4} + 3.34 \times 10^{-6} \times 350\ \mathrm{K}) = V_0(4.2 \times 10^{-4} + 11.7 \times 10^{-4}\ \mathrm{K}) = 1.59 \times 10^{-3}\ \mathrm{K}V_0$

Chapter 34

1. a. $\dfrac{d}{dx}(\ln 3x) = \dfrac{1}{x}$

b. $\dfrac{d}{dx}(e^{-5x}) = -5e^{-5x}$

c. $\dfrac{d}{dx}[\sin(4x - 7)] = 4\cos(4x - 7)$

d. $\dfrac{d}{dx}(\log 7x - \cos 2x) = \dfrac{d}{dx}\left(\dfrac{\ln 7x}{2.303} - \cos 2x\right) = \dfrac{1}{2.303\,x} - (-2\sin 2x) = \dfrac{1}{2.303x} + 2\sin 2x$

e. $\dfrac{d}{dx}[e^{-x} + \sin(3x + 2) + \ln 9x] = -e^{-x} + 3\cos(3x + 2) + \dfrac{1}{x}$

2. $\log \gamma_{\pm} = -Az_+z_-\sqrt{I} = -Az_+z_-I^{1/2}$
$\dfrac{d\log \gamma_{\pm}}{dI} = -Az_+z_-\left(\dfrac{1}{2}I^{-1/2}\right) = \dfrac{-\frac{1}{2}Az_+z_-}{I^{1/2}} = -\dfrac{Az_+z_-}{2\sqrt{I}}$

3. $\lambda = \dfrac{2d}{n}\sin\theta, \quad \dfrac{d\lambda}{d\theta} = \dfrac{2d}{n}\cos\theta$

4. $\ln k = \ln A - \dfrac{E_a}{RT} = \ln A - \dfrac{E_a}{R}T^{-1}, \quad \dfrac{d\ln k}{dT} = -\dfrac{E_a}{R}(-T^{-2}) = \dfrac{E_a}{RT^2}$

5. $\dfrac{dc}{dt} = c_0(-ke^{-kt}) = -kc_0e^{-kt}$

Chapter 35

1. a. $\frac{d}{dx}(4x^5) = 4 \times 5x^{5-1} = 20x^4$

b. $\frac{d}{dx}(x^3 - x^2 + x - 9) = 3x^2 - 2x + 1$

c. $\frac{d}{dx}(3\ln x - 4\sin 2x) = \frac{3}{x} - 4 \times 2\cos 2x = \frac{3}{x} - 8\cos 2x$

d. $\frac{d}{dx}(6x - e^{-3x} + \ln 5x) = 6 - (-3)e^{-3x} + \frac{1}{x} = 6 + 3e^{-3x} + \frac{1}{x}$

e. $\frac{d}{dx}\left(\frac{5}{x^3} - 2x + \ln 8x\right) = \frac{d}{dx}(5x^{-3} - 2x + \ln 8x) = 5 \times (-3x^{-4}) - 2 + \frac{1}{x} = -15x^{-4} - 2 + \frac{1}{x}$

$= -\frac{15}{x^4} - 2 + \frac{1}{x}$

2. a. $\frac{d}{dx}(x\ln x) = \ln x\frac{d}{dx}(x) + x\frac{d}{dx}(\ln x) = \ln x \times 1 + x \times \frac{1}{x} = \ln x + 1$

b. $\frac{d}{dx}(x^2e^{-2x}) = e^{-2x}\frac{d}{dx}(x^2) + x^2\frac{d}{dx}(e^{-2x}) = e^{-2x} \times 2x + x^2 \times (-2e^{-2x}) = 2xe^{-2x} - 2x^2e^{-2x}$

$= 2xe^{-2x}(1 - x)$

c. $\frac{d}{dx}(3x\sin 2x) = 3\left[\sin 2x\frac{d}{dx}(x) + x\frac{d}{dx}(\sin 2x)\right] = 3(\sin 2x \times 1 + x \times 2\cos 2x) = 3(\sin 2x + 2x\cos 2x)$

d. $\frac{d}{dx}(4xe^x + x) = 4\left[e^x\frac{d}{dx}(x) + x\frac{d}{dx}(e^x)\right] + \frac{d}{dx}(x) = 4(e^x \times 1 + xe^x) + 1 = 4e^x + 4xe^x + 1$

e. $\frac{d}{dx}(7x^2\cos 4x + xe^x) = 7\left[\cos 4x\frac{d}{dx}(x^2) + x^2\frac{d}{dx}(\cos 4x)\right] + \left[e^x\frac{d}{dx}(x) + x\frac{d}{dx}(e^x)\right]$

$= 7\left[2x\cos 4x + x^2(-4\sin 4x)\right] + (e^x \times 1 + xe^x) = 14x\cos 4x - 28x^2\sin 4x + e^x + xe^x$

3. a. $\frac{d}{dx}\left(\frac{x}{\ln x}\right) = \frac{\ln x\frac{d}{dx}(x) - x\frac{d}{dx}(\ln x)}{(\ln x)^2} = \frac{\ln x \times 1 - x \times \frac{1}{x}}{(\ln x)^2} = \frac{\ln x - 1}{(\ln x)^2}$

b. $\frac{d}{dx}\left(\frac{\sin x}{x^2}\right) = \frac{x^2\frac{d}{dx}(\sin x) - \sin x\frac{d}{dx}(x^2)}{(x^2)^2} = \frac{x^2\cos x - \sin x \times 2x}{x^4} = \frac{x^2\cos x - 2x\sin x}{x^4}$

$= \frac{x\cos x - 2\sin x}{x^3}$

c. $\frac{d}{dx}\left(\frac{e^x}{\ln 2x}\right) = \frac{\ln 2x\frac{d}{dx}(e^x) - e^x\frac{d}{dx}(\ln 2x)}{(\ln 2x)^2} = \frac{\ln 2x \times e^x - e^x \times \frac{1}{x}}{(\ln 2x)^2} = \frac{e^x\ln 2x - \frac{e^x}{x}}{(\ln 2x)^2} = \frac{e^x}{\ln 2x} - \frac{e^x}{x(\ln 2x)^2}$

d. $\frac{d}{dx}\left(\frac{\sin x}{\cos 3x}\right) = \frac{\cos 3x\frac{d}{dx}(\sin x) - \sin x\frac{d}{dx}(\cos 3x)}{(\cos 3x)^2} = \frac{\cos 3x \times \cos x - \sin x \times (-3\sin 3x)}{(\cos 3x)^2}$

$= \frac{\cos x\cos 3x + 3\sin x\sin 3x}{(\cos 3x)^2}$

e. $\frac{d}{dx}\left(\frac{x\ln x}{\sin x}\right) = \frac{\sin x\frac{d}{dx}(x\ln x) - x\ln x\frac{d}{dx}(\sin x)}{\sin^2 x} = \frac{\sin x\left[\ln x\frac{d}{dx}(x) + x\frac{d}{dx}(\ln x)\right] - x\ln x\cos x}{\sin^2 x}$

$= \frac{\sin x\left[\ln x \times 1 + x \times \frac{1}{x}\right] - x\ln x\cos x}{\sin^2 x} = \frac{\sin x\left[\ln x + 1\right] - x\ln x\cos x}{\sin^2 x}$

$= \frac{\sin x\ln x + \sin x - x\ln x\cos x}{\sin^2 x}$

4. Setting $x = \Lambda/\Lambda_0$ gives $K = cx/(1 - x)$.

$$\frac{dK}{dx} = \frac{c\left[(1 - x)\frac{d}{dx}(x) - x\frac{d}{dx}(1 - x)\right]}{(1 - x)^2} = \frac{c\left[(1 - x) \times 1 - x \times (-1)\right]}{(1 - x)^2} = \frac{c(1 - x + x)}{(1 - x)^2} = \frac{c}{(1 - x)^2}$$

so that

$$\frac{\mathrm{d}K}{\mathrm{d}(\Lambda/\Lambda_0)} = \frac{c}{\left(1-\frac{\Lambda}{\Lambda_0}\right)^2}$$

5. $\dfrac{\mathrm{d}}{\mathrm{d}n_i}(n_i \ln n_i - n_i) = \ln n_i \dfrac{\mathrm{d}}{\mathrm{d}n_i}(n_i) - n_i \dfrac{\mathrm{d}}{\mathrm{d}n_i}(\ln n_i) - \dfrac{\mathrm{d}}{\mathrm{d}n_i}(n_i) = \ln n_i \times 1 + n_i \times \dfrac{1}{n_i} - 1 = \ln n_i + 1 - 1 = \ln n_i$

Chapter 36

1. a. $\dfrac{\mathrm{d}f(x)}{\mathrm{d}x} = 12x^2 - 6x + 1, \quad \dfrac{\mathrm{d}^2 f(x)}{\mathrm{d}x^2} = 24x - 6$

   b. $\dfrac{\mathrm{d}f(x)}{\mathrm{d}x} = 24x^3 - 6x, \quad \dfrac{\mathrm{d}^2 f(x)}{\mathrm{d}x^2} = 72x^2 - 6$

   c. $\dfrac{\mathrm{d}f(x)}{\mathrm{d}x} = 18x + 3, \quad \dfrac{\mathrm{d}^2 f(x)}{\mathrm{d}x^2} = 18$

2. a. $\dfrac{\mathrm{d}g(y)}{\mathrm{d}y} = \dfrac{1}{y} = y^{-1}, \quad \dfrac{\mathrm{d}^2 g(y)}{\mathrm{d}y^2} = -y^{-2} = -\dfrac{1}{y^2}$

   b. $\dfrac{\mathrm{d}g(y)}{\mathrm{d}y} = (-4) \times (2)e^{-4y} = -8e^{-4y}, \quad \dfrac{\mathrm{d}^2 g(y)}{\mathrm{d}y^2} = -8 \times (-4e^{-4y}) = 32e^{-4y}$

   c. $\dfrac{\mathrm{d}g(y)}{\mathrm{d}y} = \dfrac{1}{y} + 2e^{2y} = y^{-1} + 2e^{2y}, \quad \dfrac{\mathrm{d}^2 g(y)}{\mathrm{d}y^2} = -y^{-2} + 2 \times 2e^{2y} = -\dfrac{1}{y^2} + 4e^{2y}$

3. a. $\dfrac{\mathrm{d}h(z)}{\mathrm{d}z} = 3\cos 3z, \quad \dfrac{\mathrm{d}^2 h(z)}{\mathrm{d}z^2} = 3(-3\sin 3z) = -9\sin 3z$

   b. $\dfrac{\mathrm{d}h(z)}{\mathrm{d}z} = -4\sin(4z+1), \quad \dfrac{\mathrm{d}^2 h(z)}{\mathrm{d}z^2} = -4\,[4\cos(4z+1)] = -16\cos(4z+1)$

   c. $\dfrac{\mathrm{d}h(z)}{\mathrm{d}z} = 2\cos 2z - 2\sin 2z, \quad \dfrac{\mathrm{d}^2 h(z)}{\mathrm{d}z^2} = 2(-2\sin 2z) - 2(2\cos 2z) = -4\sin 2z - 4\cos 2z$

4. $\dfrac{\mathrm{d}\Psi}{\mathrm{d}r} = \left(\dfrac{1}{\pi}\right)^{1/2}\left(\dfrac{1}{a_0}\right)^{3/2}\dfrac{\mathrm{d}}{\mathrm{d}r}\left(e^{-r/a_0}\right) = \left(\dfrac{1}{\pi}\right)^{1/2}\left(\dfrac{1}{a_0}\right)^{3/2}\left(-\dfrac{1}{a_0}\right)e^{-r/a_0} = -\left(\dfrac{1}{\pi}\right)^{1/2}\left(\dfrac{1}{a_0}\right)^{5/2}e^{-r/a_0}$

   $\dfrac{\mathrm{d}^2\Psi}{\mathrm{d}r^2} = -\left(\dfrac{1}{\pi}\right)^{1/2}\left(\dfrac{1}{a_0}\right)^{5/2}\dfrac{\mathrm{d}}{\mathrm{d}r}\left(e^{-r/a_0}\right) = -\left(\dfrac{1}{\pi}\right)^{1/2}\left(\dfrac{1}{a_0}\right)^{5/2}\left(-\dfrac{1}{a_0}\right)e^{-r/a_0}$

   $= \left(\dfrac{1}{\pi}\right)^{1/2}\left(\dfrac{1}{a_0}\right)^{7/2}e^{-r/a_0}$

5. $\dfrac{\mathrm{d}V}{\mathrm{d}n} = b + \dfrac{3}{2}cn^{1/2} + 2en, \quad \dfrac{\mathrm{d}^2 V}{\mathrm{d}n^2} = \dfrac{3}{4}cn^{-1/2} + 2e = \dfrac{3c}{4\sqrt{n}} + 2e$

   When $n = 0.25$ mol,

   $\dfrac{\mathrm{d}^2 V}{\mathrm{d}n^2} = \dfrac{3 \times 1.7738\ \text{cm}^3\ \text{mol}^{-1}}{4 \times (0.25\ \text{mol})^{1/2}} + 2 \times 0.1194\ \text{cm}^3\ \text{mol}^{-2} = (2.6607 + 0.2388)\ \text{cm}^3\ \text{mol}^{-2} = 2.8995\ \text{cm}^3\ \text{mol}^{-2}$

Chapter 37

1. $\dfrac{\mathrm{d}f(x)}{\mathrm{d}x} = 6x - 6 = 6(x-1) = 0, \ x = 1$

2. $\dfrac{\mathrm{d}g(y)}{\mathrm{d}y} = \dfrac{1}{y} - 4y = 0, \ \dfrac{1}{y} = 4y, \ 4y^2 = 1, \ y^2 = \dfrac{1}{4}, \ y = \pm\dfrac{1}{2}$

3. $\dfrac{df(x)}{dx} = 12x^2 - 12x = 12x(x-1) = 0,\ x = 0 \text{ or } x = 1.\ \dfrac{d^2f(x)}{dx^2} = 24x - 12.$

When $x = 0$, $\dfrac{d^2f(x)}{dx^2} = -12$, i.e. maximum; when $x = 1$, $\dfrac{d^2f(x)}{dx^2} = 12$, i.e. minimum.

4. $\dfrac{df(x)}{dx} = 3\cos(3x-5) = 0$

$3x - 5 = \pi/2, 3\pi/2, 5\pi/2, 7\pi/2$; $3x - 5 = 1.571, 4.713, 7.855, 10.997$; $3x = 6.571, 9.713, 12.855, 15.997$; $x = 2.190, 3.238, 4.285, 5.332$

5. $V(r) = -Ar^{-1} + Br^{-6},\ \dfrac{dV(r)}{dr} = Ar^{-2} + -6Br^{-7} = \dfrac{A}{r^2} - \dfrac{6B}{r^7} = \dfrac{1}{r^2}\left(A - \dfrac{6B}{r^5}\right),$

$\dfrac{dV(r)}{dr} = 0 \text{ when } \dfrac{6B}{r^5} = A,\ r^5 = \dfrac{6B}{A}$

$$\frac{d^2V(r)}{dr^2} = -2Ar^{-3} + 42Br^{-8} = -\frac{2A}{r^3} + \frac{42B}{r^8} = \frac{1}{r^3}\left(-2A + \frac{42B}{r^5}\right)$$

$$= \frac{1}{r^3}\left(-2A + \frac{42B}{6B} \times A\right) = \frac{1}{r^3}(-2A + 7A) = \frac{5A}{r^3}$$

Since this is positive, the stationary point is a minimum.

6. $\dfrac{dE}{dZ} = \dfrac{e^2}{a_0}\left(2Z - \dfrac{27}{8}\right) = 0,\ 2Z = \dfrac{27}{8},\ Z = \dfrac{27}{16}$

$\dfrac{d^2E}{dZ^2} = \dfrac{e^2}{a_0} \times 2$

Since this is positive, the stationary point is a minimum.

$$E = \frac{e^2}{a_0}\left[\left(\frac{27}{16}\right)^2 - \left(\frac{27}{8} \times \frac{27}{16}\right)\right] = \frac{e^2}{a_0}(2.848 - 5.695) = -2.847\frac{e^2}{a_0}$$

Chapter 38

1. $f(x, y) = (5y)x^2 - (3y)x - (4y^2)x,\ \left(\dfrac{\partial f}{\partial x}\right)_y = 10xy - 3y - 4y^2$

$f(x, y) = (5x^2)y - (3x)y - (4x)y^2,\ \left(\dfrac{\partial f}{\partial y}\right)_x = 5x^2 - 3x - 8xy$

2. $g(x, y) = x^2 + \ln x - \ln y,\ \left(\dfrac{\partial g}{\partial x}\right)_y = 2x + \dfrac{1}{x},\ \left(\dfrac{\partial g}{\partial y}\right)_x = -\dfrac{1}{y}$

3. $\left(\dfrac{\partial h}{\partial r}\right)_s = -e^{-r} + \cos(r + 2s),\ \left(\dfrac{\partial h}{\partial s}\right)_r = 2\cos(r + 2s)$

4. $\left(\dfrac{\partial p}{\partial T}\right)_V = \dfrac{nR}{V},\ \left(\dfrac{\partial p}{\partial V}\right)_T = -\dfrac{nRT}{V^2}$

$$dp = \left(\frac{\partial p}{\partial T}\right)_V dT + \left(\frac{\partial p}{\partial V}\right)_T = \frac{nR}{V}dT - \frac{nRT}{V^2}dV$$

5. $\left(\dfrac{\partial \Delta S}{\partial x_1}\right)_{x_2} = -R\left(\ln x_1 + \dfrac{1}{x_1} \times x_1\right) = -R(\ln x_1 + 1),\ \left(\dfrac{\partial \Delta S}{\partial x_2}\right)_{x_1} = -R(\ln x_2 + 1)$

$$d\Delta S = \left(\frac{\partial \Delta S}{\partial x_1}\right)dx_1 + \left(\frac{\partial \Delta S}{\partial x_2}\right)dx_2 = -R(\ln x_1 + 1)\,dx_1 - R(\ln x_2 + 1)\,dx_2$$

Chapter 39

1. $\int_1^4 x^3 \mathrm{d}x = \left[\frac{x^4}{4}\right]_1^4 = \frac{4^4 - 1^4}{4} = \frac{256-1}{4} = \frac{255}{4}$

2. $\int_{-2}^{2} \frac{1}{x^2}\mathrm{d}x = \left[-\frac{1}{x}\right]_{-2}^{2} = -\frac{1}{2} - \left(-\frac{1}{-2}\right) = -\frac{1}{2} - \frac{1}{2} = -1$

3. $\int_0^{2\pi} \sin 2x \, \mathrm{d}x = \left[-\frac{1}{2}\cos 2x\right]_0^{2\pi} = \left[-\frac{1}{2}\cos 4\pi\right] - \left[-\frac{1}{2}\cos 0\right] = \left[-\frac{1}{2}\times 1\right] - \left[-\frac{1}{2}\times 1\right] = -\frac{1}{2} + \frac{1}{2} = 0$

4. $\int_{c_0}^{c} \mathrm{d}c = -k\int_0^t \mathrm{d}t,\ [c]_{c_0}^{c} = -k\,[t]_0^t,\ c - c_0 = -k(t-0),\ c - c_0 = -kt$

5. $\int_{p_0}^{p} \frac{\mathrm{d}p}{p} = [\ln p]_{p_0}^{p} = \ln p - \ln p_0 = \ln\left(\frac{p}{p_0}\right),\ \int_0^z \mathrm{d}z = [z]_0^z = z - 0 = z$

   $\ln\left(\frac{p}{p_0}\right) = -\frac{mgz}{RT}$

Chapter 40

1. a. $\int x^9 \mathrm{d}x = \frac{x^{10}}{10} + C$

   b. $\int 3x^{-6}\mathrm{d}x = \frac{3x^{-5}}{-5} + C = -\frac{3}{5x^5} + C$

   c. $\int \frac{2}{3x}\mathrm{d}x = \frac{2}{3}\int \frac{\mathrm{d}x}{x} = \frac{2}{3}\ln x + C$

   d. $\int 2e^{-5x}\mathrm{d}x = \frac{2e^{-5x}}{-5} + C = -\frac{2}{5e^{5x}} + C$

   e. $\int \cos 3x \, \mathrm{d}x = \frac{\sin 3x}{3} + C$

2. a. $\int_0^6 (5x^3 - 2x^2 + x + 6)\,\mathrm{d}x = \left[\frac{5x^4}{4} - \frac{2x^3}{3} + \frac{x^2}{2} + 6x\right]_0^6$

   $= \left[\frac{5\times 6^4}{4} - \frac{2\times 6^3}{3} + \frac{6^2}{2} + 6\times 6\right] - [0] = \frac{5\times 1296}{4} - \frac{2\times 216}{3} + \frac{36}{2} + 36$

   $= 1620 - 144 + 18 + 36 = 1530$

   b. $\int_{-1}^{3} (x^2 - 2x + 1)\,\mathrm{d}x = \left[\frac{x^3}{3} - \frac{2x^2}{2} + x\right]_{-1}^{3} = \left[\frac{x^3}{3} - x^2 + x\right]_{-1}^{3}$

   $= \left[\frac{3^3}{3} - 3^2 + 3\right] - \left[\frac{(-1)^3}{3} - (-1)^2 + (-1)\right] = [9 - 9 + 3] - \left[-\frac{1}{3} - 1 - 1\right] = 3 + \frac{1}{3} + 2 = 5 + \frac{1}{3} = \frac{16}{3}$

   c. $\int_{-4}^{0} (3x^4 - 4x^2 - 7)\,\mathrm{d}x = \left[\frac{3x^5}{5} - \frac{4x^3}{3} - 7x\right]_{-4}^{0}$

   $= [0] - \left[\frac{3\times(-4)^5}{5} - \frac{4\times(-4)^3}{3} - 7\times(-4)\right] = 0 - \left[\frac{-3\times 1024}{5} + \frac{256}{3} + 28\right]$

   $= -\left[-\frac{3072}{5} + \frac{256}{3} + 28\right] = -\left[\frac{-9216 + 1280 + 420}{15}\right] = \frac{7516}{15} = 501$

3. a. $\int_1^5 \frac{1}{2x}\mathrm{d}x = \frac{1}{2}\int_1^5 \frac{\mathrm{d}x}{x} = \frac{1}{2}[\ln x]_1^5 = \frac{1}{2}[1.609 - 0] = 0.805$

   b. $\int_0^1 e^{3x}\mathrm{d}x = \left[\frac{e^{3x}}{3}\right]_0^1 = \frac{e^3}{3} - \frac{e^0}{3} = \frac{20.08}{3} - \frac{1}{3} = \frac{19.08}{3} = 6.36$

c. $\int_0^{\pi} \sin 4x \, \mathrm{d}x = \left[\frac{-\cos 4x}{4}\right]_0^{\pi} = \left[\frac{-\cos 4\pi}{4}\right] - \left[\frac{-\cos 0}{4}\right] = -\frac{1}{4} + \frac{1}{4} = 0$

4. $\Delta H = \int_{T_1}^{T_2} \left(a + bT + cT^{-2}\right) \mathrm{d}T = \left[aT + \frac{bT^2}{2} + \frac{cT^{-1}}{-1}\right]_{T_1}^{T_2} = \left[aT + \frac{bT^2}{2} - \frac{c}{T}\right]_{T_1}^{T_2}$

$= a(T_2 - T_1) + \frac{b}{2}\left(T_2^2 - T_1^2\right) - \frac{c}{T_2 - T_1}$

$= 28.58\,\mathrm{J\,K^{-1}\,mol^{-1}} \times (318 - 298)\,\mathrm{K} + \frac{3.76 \times 10^{-3}\,\mathrm{J\,K^{-2}\,mol^{-1}}}{2} \times (318^2 - 298^2)\,\mathrm{K}^2 + \frac{5.0 \times 10^4\,\mathrm{J\,K\,mol^{-1}}}{(318 - 298)\,\mathrm{K}}$

$= 28.58 \times 20\,\mathrm{J\,mol^{-1}} + 1.88 \times 10^{-3} \times (101\,124 - 88\,804)\,\mathrm{J\,mol^{-1}} + \frac{5.0 \times 10^4\,\mathrm{J\,mol^{-1}}}{20}$

$= 571.60\,\mathrm{J\,mol^{-1}} + 1.88 \times 10^{-3} \times 12\,320\,\mathrm{J\,mol^{-1}} + 2500\,\mathrm{J\,mol^{-1}}$

$= (571.60 + 23.16 + 2500)\,\mathrm{J\,mol^{-1}} = 3095\,\mathrm{J\,mol^{-1}} = 3.10\,\mathrm{kJ\,mol^{-1}}$

5. $\int_{K_1}^{K_2} \frac{\mathrm{d}K}{K} = [\ln K]_{K_1}^{K_2} = \ln K_2 - \ln K_1 = \ln\left(\frac{K_2}{K_1}\right)$

$\int_{T_1}^{T_2} \frac{\mathrm{d}T}{T^2} = \int_{T_1}^{T_2} T^{-2}\mathrm{d}T = \left[\frac{T^{-1}}{-1}\right]_{T_1}^{T_2} = \left[-\frac{1}{T}\right]_{T_1}^{T_2} = -\frac{1}{T_2} - \left(-\frac{1}{T_1}\right) = \frac{1}{T_1} - \frac{1}{T_2}$

$\ln\left(\frac{K_2}{K_1}\right) = \frac{\Delta H^{\ominus}}{R}\left(\frac{1}{T_1} - \frac{1}{T_2}\right)$

Chapter 41

1. a. $u = x,\ \frac{\mathrm{d}u}{\mathrm{d}x} = 1,\ \frac{\mathrm{d}v}{\mathrm{d}x} = \sin x,\ v = -\cos x$

$$\int x \sin x \, \mathrm{d}x = x(-\cos x) - \int (-\cos x) \times 1 \, \mathrm{d}x = -x\cos x + \int \cos x \, \mathrm{d}x = -x\cos x + \sin x + C$$

b. $u = x,\ \frac{\mathrm{d}u}{\mathrm{d}x} = 1,\ \frac{\mathrm{d}v}{\mathrm{d}x} = e^x,\ v = e^x$

$$\int xe^x \mathrm{d}x = xe^x - \int e^x \mathrm{d}x = xe^x - e^x + C$$

c. $u = \ln x,\ \frac{\mathrm{d}u}{\mathrm{d}x} = \frac{1}{x},\ \frac{\mathrm{d}v}{\mathrm{d}x} = x,\ v = \frac{x^2}{2}$

$$\int x \ln x \, \mathrm{d}x = \frac{x^2}{2}\ln x - \int \frac{x^2}{2} \times \frac{1}{x}\mathrm{d}x = \frac{x^2}{2}\ln x - \frac{1}{2}\int x \, \mathrm{d}x = \frac{x^2}{2}\ln x - \frac{1}{2} \times \frac{x^2}{2} + C = \frac{x^2}{2}\ln x - \frac{x^2}{4} + C$$

d. $u = x,\ \frac{\mathrm{d}u}{\mathrm{d}x} = 1,\ \frac{\mathrm{d}v}{\mathrm{d}x} = e^{3x},\ v = \frac{e^{3x}}{3}$

$$\int xe^{3x}\mathrm{d}x = x\frac{e^{3x}}{3} - \int \frac{e^{3x}}{3}\mathrm{d}x = x\frac{e^{3x}}{3} - \frac{e^{3x}}{9} + C$$

$$\int_1^3 xe^{3x}\mathrm{d}x = \left[\frac{xe^{3x}}{3} - \frac{e^{3x}}{9}\right]_1^3 = \left[\frac{3e^9}{3} - \frac{e^9}{9}\right] - \left[\frac{e^3}{3} - \frac{e^3}{9}\right] = [8103 - 900] - [6.70 - 2.23]$$

$= 7203 - 4.47 = 7199$

e. $u = 3x,\ \frac{\mathrm{d}u}{\mathrm{d}x} = 3,\ \frac{\mathrm{d}v}{\mathrm{d}x} = \cos 2x,\ v = \frac{\sin 2x}{2}$

$$\int 3x \cos 2x \, \mathrm{d}x = \frac{3x \sin 2x}{2} - \int \frac{3 \sin 2x}{2} = \frac{3x \sin 2x}{2} - \frac{3}{2}\frac{(-\cos 2x)}{2} = \frac{3x \sin 2x}{2} + \frac{3\cos 2x}{4} + C$$

$$\int_0^{\pi/2} \cos 2x \, \mathrm{d}x = \left[\frac{3x \sin 2x}{2} + \frac{3\cos 2x}{4}\right]_0^{\pi/2} = \left[\frac{3 \times \frac{\pi}{2}\sin \pi}{2} + \frac{3\cos \pi}{4}\right] - \left[0 + \frac{3}{4}\cos 0\right]$$

$$= \frac{3\pi}{4} \times 0 + \frac{3}{4} \times (-1) - 0 - \frac{3}{4} \times 1 = -\frac{3}{4} - \frac{3}{4} = -\frac{6}{4} = -\frac{3}{2}$$

2. a. $u = x - 2,\ \frac{du}{dx} = 1$
When $x = -1$, $u = -3$; when $x = 4$, $u = 2$.

$$\int_{-3}^{2} u^6 du = \left[\frac{u^7}{7}\right]_{-3}^{2} = \frac{2^7 - (-3)^7}{7} = \frac{128 + 2187}{7} = \frac{2315}{7} = 331$$

b. $u = 4x + 1,\ \frac{du}{dx} = 4$
When $x = -\pi$, $u = -4\pi + 1 = -11.56$; when $x = \pi/2$, $u = 2\pi + 1 = 7.28$.

$$\int_{-11.56}^{7.28} \sin u \frac{du}{4} = \frac{1}{4}\int_{-11.56}^{7.28} \sin u\, du = \frac{1}{4}[-\cos u]_{-11.56}^{7.28} = \frac{1}{4}[-\cos 7.28 - (-\cos(-11.56))]$$

$$= \frac{1}{4}[-0.540 + 0.540] = 0.0$$

c. $u = x^2,\ \frac{du}{dx} = 2x,\ dx = \frac{du}{2x}$
When $x = 0$, $u = 0$; when $x = 2$, $u = 4$.

$$\int_0^4 3xe^u \frac{du}{2x} = \int_0^4 \frac{3}{2}e^u\, du = \frac{3}{2}[e^u]_0^4 = \frac{3}{2}[e^4 - e^0] = \frac{3}{2}[54.60 - 1] = 80.4$$

3. a. $\frac{3x}{(x+1)(x-4)} = \frac{A}{x+1} + \frac{B}{x-4} = \frac{A(x-4) + B(x+1)}{(x+1)(x-4)}$
$3x = A(x-4) + B(x+1)$
When $x = 4$, $3 \times 4 = B(4+1)$, $12 = 5B$, $B = 12/5$.
When $x = -1$, $3 \times (-1) = A(-1-4)$, $-3 = -5A$, $A = -3/-5 = 3/5$.

$$\int_5^7 \left(\frac{dx}{2(x+1)} + \frac{12\, dx}{5(x-4)}\right) = \left[\frac{3}{5}\ln(x+1) + \frac{12}{5}\ln(x-4)\right]_5^7$$

$$= \left[\frac{3}{5}\ln 8 + \frac{12}{5}\ln 3\right] - \left[\frac{3}{5}\ln 6 + \frac{12}{5}\ln 1\right]$$

$$= \left[\frac{3}{5} \times 2.08 + \frac{12}{5} \times 1.10\right] - \left[\frac{3}{5} \times 1.79 + \frac{12}{5} \times 0\right]$$

$$= [1.25 + 2.64] - [1.08] = 2.81$$

b. $\frac{2}{x(x+1)} = \frac{A}{x} + \frac{B}{x+1} = \frac{A(x+1) + Bx}{x(x+1)}$
$2 = \mathrm{A}(x+1) + Bx$
When $x = -1$, $2 = B \times (-1)$, $B = -2$; when $x = 0$, $2 = A$.

$$\int \frac{2}{x(x+1)}dx = \int \frac{2}{x}dx - \int \frac{2}{x+1}dx = 2\ln x - 2\ln(x+1) + C$$

c. $\frac{2x-3}{(x+5)(x-2)} = \frac{A}{x+5} + \frac{B}{x-2} = \frac{A(x-2) + B(x+5)}{(x+5)(x-2)}$
$2x - 3 = A(x-2) + B(x+5)$
When $x = 2$, $2 \times 2 - 3 = B(2+5)$, $1 = 7B$, $B = 1/7$.
When $x = -5$, $2 \times (-5) - 3 = A(-5-2)$, $-13 = -7A$, $A = 13/7$.

$$\int_5^{10} \frac{13}{7(x+5)}dx + \int_5^{10} \frac{1}{7(x-2)}dx = \left[\frac{13}{7}\ln(x+5) + \frac{1}{7}\ln(x-2)\right]_5^{10}$$

$$= \frac{13}{7}\ln 15 + \frac{1}{7}\ln 8 - \frac{13}{7}\ln 10 - \frac{1}{7}\ln 3 = \frac{13}{7} \times 2.708 + \frac{1}{7} \times 2.079 - \frac{13}{7} \times 2.303 - \frac{1}{7} \times 1.099$$

$$= 5.029 + 0.297 - 4.277 - 0.157 = 0.892$$

4. $u = r^2,\ \dfrac{du}{dr} = 2r,\ \dfrac{dv}{dr} = e^{-2r/a_0},\ v = -\dfrac{a_0}{2}e^{-2r/a_0}$

$$\int r^2 e^{-2r/a_0}\mathrm{d}r = r^2\left(-\frac{a_0}{2}\right)e^{-2r/a_0} - \int\left(-\frac{a_0}{2}\right)e^{-2r/a_0}2r\mathrm{d}r = \left(-\frac{a_0}{2}\right)r^2 e^{-2r/a_0} + a_0\int re^{-2r/a_0}\mathrm{d}r$$

Determine $\int re^{-2r/a_0}\,\mathrm{d}r$ by setting $u = r$ and $\dfrac{\mathrm{d}v}{\mathrm{d}r} = e^{-2r/a_0}$. This gives $\mathrm{d}u/\mathrm{d}r = 1$ and $v = -\frac{a_0}{2}e^{-2r/a_0}$, as before.

$$\int re^{-2r/a_0}\mathrm{d}r = r\left(-\frac{a_0}{2}\right)e^{-2r/a_0} - \int\left(-\frac{a_0}{2}\right)e^{-2r/a_0}\times 1\,\mathrm{d}r$$

$$= \left(-\frac{a_0}{2}\right)re^{-2r/a_0} + \left(\frac{a_0}{2}\right)\left(-\frac{a_0}{2}\right)e^{-2r/a_0} + C = \left(-\frac{a_0}{2}\right)re^{-2r/a_0} + \left(-\frac{a_0^2}{4}\right)e^{-2r/a_0} + C$$

Substitute to give

$$\int r^2 e^{-2r/a_0}\mathrm{d}r = \left(-\frac{a_0}{2}\right)r^2 e^{-2r/a_0} + a_0\left[\left(-\frac{a_0}{2}\right)re^{-2r/a_0} + \left(-\frac{a_0^2}{4}\right)e^{-2r/a_0}\right] + C$$

$$= a_0 e^{-2r/a_0}\left[-\frac{r^2}{2} - \frac{a_0 r}{2} - \frac{a_0^2}{4}\right] + C$$

5. $u = J(J+1) = J^2 + J,\ \dfrac{\mathrm{d}u}{\mathrm{d}J} = 2J + 1$
When $J = 0$, $u = 0$; when $J = \infty$, $u = \infty$.

$$q_r = \int_0^\infty (2J+1)e^{-uh^2/8\pi^2 IkT}\frac{\mathrm{d}u}{2J+1} = \int_0^\infty e^{-uh^2/8\pi^2 IkT}\mathrm{d}u = \left(\frac{-8\pi^2 IkT}{h^2}\right)\left[e^{-uh^2/8\pi^2 IkT}\right]_0^\infty$$

$$= \left(\frac{-8\pi^2 IkT}{h^2}\right)[0-1] = \frac{8\pi^2 IkT}{h^2}$$

Index